THE FRONTIERS COLLECTION

THE FRONTIERS COLLECTION

The books in this collection are devoted to challenging and open problems at the forefront of modern science, including related philosophical debates. In contrast to typical research monographs, however, they strive to present their topics in a manner accessible also to scientifically literate non-specialists wishing to gain insight into the deeper implications and fascinating questions involved. Taken as a whole, the series reflects the need for a fundamental and interdisciplinary approach to modern science. Furthermore, it is intended to encourage active scientists in all areas to ponder over important and perhaps controversial issues beyond their own speciality. Extending from quantum physics and relativity to entropy, consciousness and complex systems – the Frontiers Collection will inspire readers to push back the frontiers of their own knowledge.

Information and Its Role in Nature
By J. G. Roederer

Relativity and the Nature of Spacetime
By V. Petkov

Quo Vadis Quantum Mechanics?
Edited by A. C. Elitzur, S. Dolev, N. Kolenda

Life – As a Matter of Fat
The Emerging Science of Lipidomics
By O. G. Mouritsen

Quantum–Classical Analogies
By D. Dragoman and M. Dragoman

Knowledge and the World
Challenges Beyond the Science Wars
Edited by M. Carrier, J. Roggenhofer, G. Küppers, P. Blanchard

Quantum–Classical Correspondence
By A. O. Bolivar

Mind, Matter and Quantum Mechanics
By H. Stapp

Quantum Mechanics and Gravity
By M. Sachs

Extreme Events in Nature and Society
Edited by S. Albeverio, V. Jentsch, H. Kantz

The Thermodynamic Machinery of Life
By M. Kurzynski

The Emerging Physics of Consciousness
Edited by J. A. Tuszynski

Weak Links
Stabilizers of Complex Systems from Proteins to Social Networks
By P. Csermely

Mind, Matter and the Implicate Order
By P.T.I. Pylkkänen

Quantum Mechanics at the Crossroads
New Perspectives from History, Philosophy and Physics
By J. Evans, A.S. Thomdike

Particle Metaphysics
A Critical Account of Subatomic Reality
By B. Falkenburg

The Physical Basis of the Direction of Time
By H.D. Zeh

Asymmetry: The Foundation of Information
By S.J. Muller

Scott J. Muller

ASYMMETRY: THE FOUNDATION OF INFORMATION

With 33 Figures

Springer

Scott J. Muller

Bernoulli Systems
Suite 145
National Innovation Centre
Australian Technology Park
Eveleigh, NSW 1430
Australia
email: smuller@bernoullisystems.com

Series Editors:

Cover figure: Image courtesy of the Scientific Computing and Imaging Institute, University of Utah (www.sci.utah.edu).

Library of Congress Control Number: 2007922925

ISSN 1612-3018
ISBN 978-3-540-69883-8 Springer Berlin Heidelberg New York

Springer is a part of Springer Science+Business Media

springer.com

Typesetting: Data supplied by the author
Production: LE-TEX Jelonek, Schmidt & Vöckler GbR, Leipzig
Cover design: KünkelLopka, Werbeagentur GmbH, Heidelberg

Printed on acid-free paper SPIN 11783350 57/3100/YL - 5 4 3 2 1 0

Preface

Objects have the capacity to distinguish themselves from other objects and from themselves at different times. The interaction of objects, together with the process of making distinctions, results in the transfer of a quantity that we call information. Some objects are capable of distinguishing themselves in more ways than others. These objects have a greater information capacity. The quantification of how objects distinguish themselves and the relationship of this process to information is the subject of this book.

As individual needs have arisen in the fields of physics, electrical engineering and computational science, diverse theories of information have been developed to serve as conceptual instruments to advance each field. Based on the foundational Statistical Mechanical physics of Maxwell and Boltzmann, an entropic theory of information was developed by Brillouin, Szilard and Schrödinger. In the field of Communications Engineering, Shannon formulated a theory of information using an entropy analogue. In computer science a "shortest descriptor" theory of information was developed independently by Kolmogorov, Solomonoff and Chaitin.

The considerations presented in this book are an attempt to illuminate the common and essential principles of these approaches and to propose a unifying, non-semantic theory of information by demonstrating that the three current major theories listed above can be unified under the concept of asymmetry, by deriving a general equation of information through the use of the algebra of symmetry, namely Group Theory and by making a strong case for the thesis that information is grounded in asymmetry.

The book draws on examples from a number of fields including chemistry, physics, engineering and computer science to develop the notions of information and entropy and to illustrate their interrelation. The work is intended for readers with a some background in science or mathematics, but it is hoped the overarching concepts are general enough and their presentation sufficiently clear to permit the non-technical reader to follow the discussion.

Chapter 1 provides an introduction to the topic, defines the scope of the project and outlines the way forward. The technical concepts of entropy and probability are developed in Chapter 2 by surveying current theories of information. Distinguishability and its relationship to information is presented in Chapter 3 along with numerous illustrative examples. Chapter 4 introduces symmetry and Group Theory. This chapter demonstrates the connections between information, entropy and symmetry and shows how these can unify current information theories. Finally Chapter 5 summarises the project and identifies some open questions.

This book represents a first step in developing a theory that may serve as a general tool for a number of disciplines. I hope that it will be of some use to researchers in fields that require the development of informatic metrics or are concerned with the dynamics of information generation or destruction. Extending this, I would like to see the group-theoretic account of information develop into an algebra of causation by the quantification of transferred information.

A large portion of this research was conducted as part of my PhD dissertation at the University of Newcastle, Australia. I would like to express my deep gratitude to Cliff Hooker and John Collier for invaluable advice and guidance and to George Willis for assistance with Group Theory, in particular Topological Groups. Early discussions with Jim Crutchfield at the Santa Fe Institute were useful in clarifying some initial ideas. I would also like to thank Chris Boucher, Ellen Watson, Jamie Pullen, Lesley Roberts and Melinda Stokes for much support and inspiration. Finally, I would also like to thank my parents, Jon and Lyal.

Sydney, April 2007 *Scott Muller*

Contents

1

Introduction

Information is a primal concept about which we have deep intuitions. It forms part of our interface to the world. Thus is seems somewhat odd that it is only in the last one hundred years or so that attempts have been made to create mathematically rigorous definitions for information. Perhaps this is due to a tendency to cast information in an epistemological or semantic light, thus rendering the problem difficult to describe using formal analysis. Yet physical objects[1] are endowed with independent, self-descriptive capacity. They have innate discernable differences that may be employed to differentiate them from others or to differentiate one state of an object from another state. These objects vary in complexity, in the number of ways that they can distinguish themselves.

Recent attempts to quantify information have come at the problem with the perspective and toolkits of several specific research areas. As individual needs have arisen in such fields as physics, electrical engineering and computational science, theories of information have been developed to serve as conceptual instruments to advance that field. These theories were not developed totally in isolation. For example, Shannon [72] in communications engineering was aware of the work done by Boltzmann, and Chaitin [21], in computational science, was aware of Shannon's work. Certain concepts, such as the use of the frequency concept of probability, are shared by different information theories, and some terminology, such as 'entropy', is used in common, though often with divergent meanings. However for the most part these theories of information, while ostensibly describing the same thing, were developed for specific local needs and only partially overlap in scope.

[1] This can also include representations of abstract objects such as numbers and laws.

The resulting situation is a little like the old joke about the blind men who were asked to describe an elephant: each felt a different part of it and each came up with a different account. It is not that their individual descriptions were incorrect; it is just that they failed to establish the full picture. I believe that, in the same way, the current theories of information do not directly address the underlying essence of information. It is my intention here to start describing the whole elephant; to begin to give a comprehensive definition of information that reconciles and hopefully extends the theories developed to date.

In the context of this discussion, I take information to be an objective property of an object that exists independently of an observer, a non-conservative quantity that can be created or destroyed and that is capable of physical work. I assume these things at the outset and will also provide demonstrations to support them through the course of my argument.

As my starting point, I take my lead from two theses. The first, promoted by Collier [24] and others, states that information originates in the breaking of symmetries. The other is E.T. Jaynes' Principle of Maximum Entropy [40]. The symmetry breaking notion leads me to postulate that information is a way of abstractly representing asymmetries. The Maximum Entropy Principle requires that all the information in a system be accounted for by the removal of non-uniform (read asymmetric) distributions of microstates until an equiprobable description is attained for the system. These two approaches, both heavily grounded in asymmetry, lead me to believe that if one is to quantify information, one must quantify asymmetries.

In this book I have three primarily goals. The first is to demonstrate that the three current major theories – the Thermodynamic/Statistical Mechanics Account, Communication Theory and Algorithmic Information Theory – can be unified under the concept of asymmetry. The second is to derive a general equation of information through the use of the algebra of symmetry, namely Group Theory. And finally I hope to make a strong case for the thesis that information is grounded in asymmetry.

Once developed, this approach might then be used by the three fields mentioned above to extend research into information itself. Moreover, because it provides an algebra of information, it can be a valuable tool for the analysis of physical systems in disparate scientific fields.

1.1 Structure

Following this introduction, Chapter 2 is a review of the aforementioned current theories of information. The first port of call (Section 2.2.1) is the Thermodynamic Theory of Information. Since the relationship between entropy and information has been well established, the section examines in some detail the history of entropic theory, the Second Law of Thermodynamics and the combinometric nature of entropy under the paradigm of Statistical Mechanics. This leads to a detailed examination of Maxwell's Demon: a thought experiment that ostensibly violates the Second Law and demonstrates the relationship between thermodynamic or physical entropy and information. This review of the Thermodynamic/Statistical Mechanics Theory of Information draws out four key concepts: the combinometric nature of entropy, the role of measurement in information systems, the role of memory in information systems and the capacity of informatic systems to do work. These are all examined in detail later in the work.

Section 2.2.2 looks briefly at Claude Shannon's contribution to the study of information, his development of a Boltzmann-like entropy theorem to quantify information capacity.

Section 2.2.3 examines the last of the three major information theories, Algorithm Information Theory. This section considers the work of Kolmogorov, Solomonoff and Chaitin, all of whom contributed to the 'shortest descriptor of a string' approach to information. Crucial to the development of their work are the notions of randomness and Turing machines. These are also studied in this section.

The general concept and specific nature of probability play an important role in all theories of information. Maxwell, Boltzmann and Shannon employ probabilistic accounts of system states. Kolmogorov stresses the importance of the frequency concept of probability. In order to develop a view of probability to use a symmetry-theory of information, Section 2.3 considers the nature of probability.

The construction of a foundational theory of information is started in Chapter 3. Commencing with a Leibnizian definition of distinguishability, the relationship between information and distinguishability is established. Based on this relationship, an objective, relational model is defined which couples an informatic object with an information gathering system. This model will serve as the infrastructure for the mathematical description of information developed in Chapter 4.

As a precursor to the development of the formal definition of information, Chapter 4 begins by examining symmetry through a brief introduction to the algebra of symmetry, Group Theory. Based on the previously constructed model of distinguishability, a formal account of information in terms of Group Theory is developed using Burnside's Lemma in Section 4.2.2. This relationship between symmetry and information is discussed at some length in Section 4.3 looking in particular at the generation of information and different types of information.

Section 4.4 considers the association of information with probability, with special interest paid to Bayes' theorem and Jaynes' Maximum Entropy Principle. Bertrand's Paradox is investigated as an example of information generated from asymmetries. The Statistical Mechanical Theory of Information is cast in the light of my analysis of information as asymmetry in Section 4.5, with attention given to the Maxwell's Demon paradox. In Section 4.6 We examine the relationship between symmetry and physical entropy and the status of the Third Law of Thermodynamics, when formulated in terms of the symmetry theory. This section also further develops the principle that information can facilitate physical work by considering Gibbs' Paradox.

The primary issues linking Algorithmic Information Theory and the asymmetry account of information centre on the notions of randomness, redundancy and compressibility. Thus these are considered in Section 4.8 by way of an example using the transcendental numbers.

Chapter 5 concludes the books and examines the need and opportunities for further work. Proof of Burnside's Lemma and worked examples used in the body of the text are provided in the Appendices.

I intend throughout this book to draw on examples and techniques from a variety of scientific fields. To avoid confusion and the possibility of losing sight of our ultimate goal, I will occasionally include signposts to summarise where we are and to indicate the direction in which we are heading.

2

Information

2.1 Scope of Information

It is prudent to initially establish the scope of what we mean by 'information'. Many contemporary philosophical theories of information are subjective in nature. Daniel Dennett , for example, maintains that information depends on the epistemic state of the receiver and as such is not independently quantifiable [29]. My understanding of information, however, is otherwise. I take information to be objective and physical, to exist independent of the observer and to be capable of producing work. Although the transfer of information from an informatic object to an external observer is bounded by the capabilities of the observer (that is, the subset of information perceived is closed by the observer), nonetheless the informational attributes of an informatic object exist independently of the existence of any observer.

What sort of information are we talking about? Information consists of any attributes that can determine, even partially, the state of an object. This may be genetic information, linguistic information, electromagnetic radiation, crystal structures, clock faces, symbolic data strings: practically anything. When I refer to 'information' in a quantitative sense, I will use the term synonymously with '*informatic capacity*'. I will labour this point somewhat. I take my definition of 'information' to be strictly non-epistemic. Though I will talk of one object O_1 "having informatic capacity with respect to" another object, O_2, the information exists independently of human apprehension. The O_2 may well be an inanimate crystal. The information is objective in the sense that it is a property of the object O_1, filtered by O_2. Information is the facility of an object to distinguish itself.

In this manner, my use of the term 'information' is strictly non-semantic. It is independent of context and content. It would, for example, treat any two 6 character strings composed of any of the 26 English letters, with no other constraints imposed, are informatically equivalent. '$f\ l\ o\ w\ e\ r$' is informatically equivalent to '$z\ y\ x\ w\ v\ u$'. However, the theory, once developed, will be capable of taking into account such semantic constraints to show how the two strings above informatically different in the context of a meta-system.

2.2 A Survey of Information Theories

Work on theories of physical information has, over the past century, arisen from three distinct, though interconnected fields: Thermodynamics and Statistical Mechanics, Communication Theory and, more recently, Algorithmic Information Theory. In each of these fields an attempt has been made to try to quantify the amount of information that is contained in a physical entity or system. Thermodynamic/Statistical Mechanics (TDSM) approaches have tried to relate a system's thermodynamic macroproperty, entropy, to the system's information content by equating information with the opposite sign of entropy: *negentropy*. Entropy, by means of Statistical Mechanics, was shown to represent a lack of information concerning the microstates of a system subject to macroconstraints. Post-war research into the burgeoning field of telecommunications during the late 1940's led to the creation of Communications Theory (also ambitiously termed "Information Theory"), in which transfer of information via channels was quantified in terms of a probability distribution of message components. A quantity that represented the reduction in uncertainty that a receiver gained on receipt of the message was found to possess a functional form similar to the entropy of Statistical Mechanics, and so was equivocally also termed entropy.

The third approach attacked the problem from a different angle. In Algorithmic Information Theory, the information content of a string representation of a system or entity is defined as the number of bits of the smallest program it takes to generate that string.[1] It has been shown that this quantification is also related to both the Statistical Mechanics and Communication Theory entropies. This section examines these three approaches and the relationship to each other in some detail.

---

[1] A string is taken to mean a sequence of symbols, usually alphanumeric characters.

2.2.1 Thermodynamic Information Theory

All modern physical theories of information make reference to a quantity known as *entropy*. The term was originally applied by the German physicist Rudolf Clausius in the middle nineteenth century to distinguish between heat source qualities in work cycles. Later work by Boltzmann provided a formal relationship between Clausius' macrolevel entropy and the microdynamics of the molecular level; this was the origin of Statistical Mechanics.

In this section we examine thermodynamic entropy, Statistical Mechanical entropy and the Second Law of Thermodynamics and their relationship with information.

Thermodynamic Entropy and the Second Law

The origin of the concept of entropy lies in the 1800s during which time rapid industrial expansion was being powered by increasingly more complex steam engines. Such engines were specific instances of a more general class of engines known as heat engines. A heat engine is defined as any device that takes heat as an energy source and produces mechanical work. The notion of entropy itself was born out of early considerations of the efficiency of heat engines.

The conversion of work to heat is a relatively simple affair. The process of friction, for example, can be analysed by considering the amount of work (W) applied and the quantity of heat generated (Q). The first law of thermodynamics tells us that the quantity of heat generated is equal to the amount of work applied: $Q = W$. That is to say that the efficiency of energy conversion is 100%. Furthermore this conversion can be carried out indefinitely. This is the First Law of Thermodynamics.

The conversion of heat to work, however, is less straightforward. The isothermal expansion of a hot gas against a piston will produce mechanical work, but eventually the pressure relative to the external pressure will drop to a point where no more work can be done. Without some sort of cyclic process whereby the system is periodically returned to its initial state after producing work, the process will not continue indefinitely.

If a cyclic process is employed, each one of the cycles consists of a number of steps in which the system interacts with the surrounding environment. The cycle of a heat engine will consist of a series of exchanges between itself and the environment where it:

- takes heat (Q_H)from a reservoir at high temperature;
- delivers heat (Q_L) to a reservoir at low temperature;
- delivers work (W) to the environment.

A schematic of the process is shown below in Fig. 2.1.

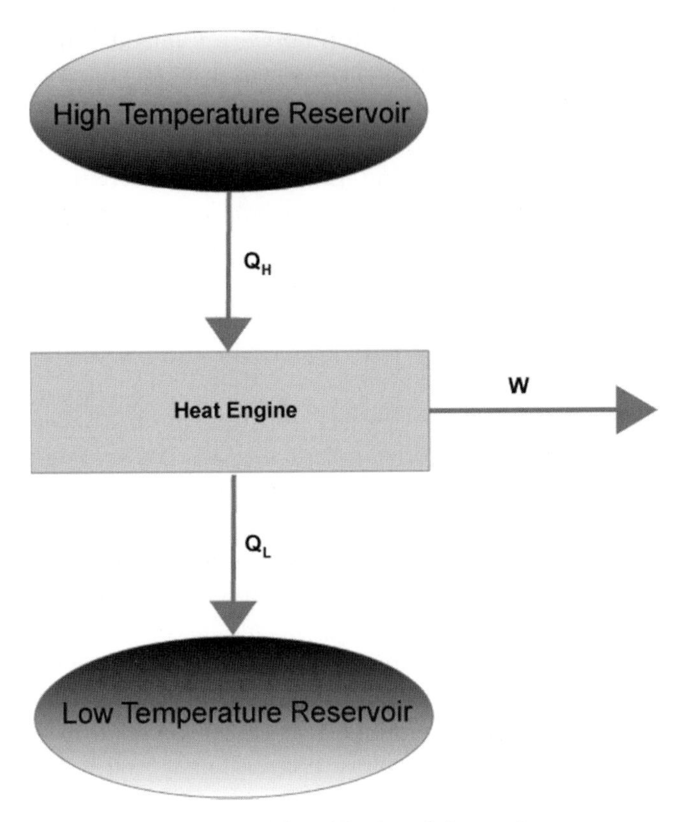

Fig. 2.1. Heat Engine Schematic

The efficiency η of such an engine is defined as work obtained per unit heat in:

$$\eta = \frac{W}{Q_H}$$

The first law of thermodynamics again tells us that, given no internal accumulation of energy, the work produced is equal to the difference between heat in and heat out:

$W = Q_H - Q_L$. Thus the efficiency equation becomes:

$$\eta = \frac{Q_H - Q_L}{Q_H}$$

or:

$$\eta = 1 - \frac{Q_L}{Q_H}$$

From this expression we can see that the efficiency of a heat engine will only be unity (i.e. 100%) if Q_L is 0, that is that there be no flow from the engine.

In 1824 a French engineer, Sadi Carnot, published "*Réflexions sur la puissance du feu, et sur les machines propre à développer cette puissance*" (Reflections on the Motive Power of Fire and on Machines Fitted to Develop that Power). Carnot showed that the most efficient engine (subsequently termed a *Carnot Engine*) is one in which all operations in the cycle are reversible. That is to say: *No engine operating between two heat reservoirs can be more efficient than a Carnot Engine operating between the same two reservoirs.* This is known as Carnot's Theorem. It should be noted that since every operation in the Carnot Engine is reversible, the whole engine could be run in reverse to create a *Carnot Refrigerator*. In this mode the same work W is performed on the engine and heat $Q_L$is absorbed from the low temperature reservoir. Heat $Q_H$is rejected to the high temperature reservoir, thus pumping heat from a low temperature to a higher temperature.

In thermodynamics, *reversibility* has a very specific meaning. A process is reversible just in case that: 1) heat flows are infinitely rapid so that the system is always in quasi-equilibrium with the environment and 2) there are no dissipative effects so the system is, in a sense, thermally frictionless. In the reversible Carnot Engine there are no friction losses or waste heat. It can be run backwards with the same absolute values of W, Q_L and Q_H to act as a heat pump. One cycle of the Carnot Engine running in normal mode followed by one cycle running in reverse (refrigerator) mode would leave the engine system (as shown in Fig. 2.2) *and the surrounding universe* completely unchanged. Carnot's Theorem states that no engine is more efficient than a Carnot engine. We can see why as follows. Imagine a candidate engine X operating between the same reservoirs shown in Fig. 2.2, taking the same heat Q_H in and depositing Q_L out and assume that the work produced is W'. Now assume that $W' > W$. If this were the case, we should be able to set aside W Joules of work from engine X to run a Carnot refrigerator between the two reservoirs and produce $W' - W$ Joules of extra work with no other effect. This is clearly impossible. At the very most $W' = W$. Here we note that W represents a cap on the amount of work that may be obtained from this heat source system. This value is independent of the design of engines. It is, as Feynman puts it, "a property of the world, not a property of a particular engine" [33].

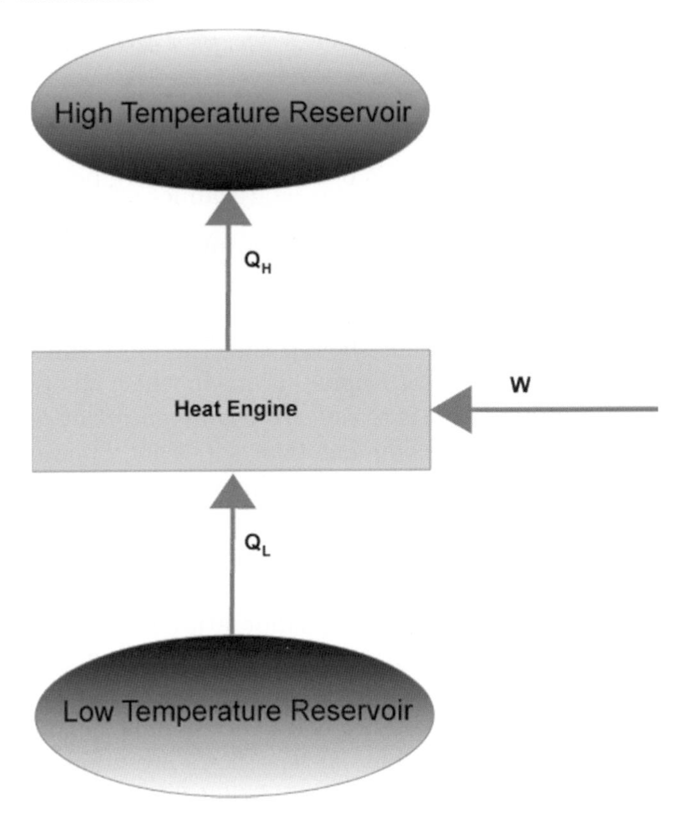

Fig. 2.2. : Carnot Refrigerator

In the real world there exists no process that operates without loss. There is no such thing as a frictionless piston. So a Carnot Engine cannot actually exist and it is this discrepancy between real world engines and Carnot's Engine that was the motivation for thought about the Second Law of Thermodynamics. It is our experience that no engine – natural or constructed – has been found to convert heat to mechanical work and deliver no waste heat. This is the basis of the Second Law of Thermodynamics and, based on empirical evidence, we assume it to be axiomatic in nature. Planck considers this and offers the following definition:

> "Since the second fundamental principle of thermodynamic is, like the first, an empirical law, we can speak of its proof only in so far as its total purport may be deduced from a single simple law of experience about which there is no doubt. We, therefore, put forward the following proposition as being given by direct experience: *It is impossible to construct an engine which will*

work in a complete cycle, and produce no effect except the raising of a weight and the cooling of a heat reservoir" [63].

The term *entropy* was introduced into the physics lexicon by Rudolf Clausius in 1865 [22]. Clausius advanced the field of thermodynamics by formalising the Second Law of Thermodynamics using methodology developed by Carnot. Clausius showed that a continuous reversible heat engine cycle could be modelled as many reversible steps which may be considered as steps in consecutive Carnot cycles. For the entire cycle consisting of j Carnot cycles, the following relationship holds true:

$$\sum_j \frac{Q_j}{T_j} = 0$$

where Q_j is the heat transferred in Carnot cycle j at temperature T_j. By taking the limit as each step size goes to 0 and j goes to infinity, an equation may be developed for a continuous reversible cycle:

$$_R\int \frac{dQ}{T} = 0$$

the R indicates that expression is true only for reversible cycles.

It follows from the preceding, that any reversible cycle may be divided in two parts: an outgoing path P_1 (from point a to point b on the cycle) and a returning path P_2 (from point b to point a), with the result that

$$_{RP_1}\int_a^b \frac{dQ}{T} =_{RP_2} - \int_b^a \frac{dQ}{T}$$

and,

$$_{RP_1}\int_a^b \frac{dQ}{T} =_{RP_2} \int_a^b \frac{dQ}{T}$$

This indicates that the quantity $_R\int_a^b \frac{dQ}{T}$ is independent of the actual reversible path from a to b. Thus there exists a thermodynamic property[2], the difference of which between a final state and an initial state is equal to the quantity $_R\int_a^b \frac{dQ}{T}$. Clausius named this property entropy and, assigning it the symbol S, defined it as follows:

$$_R\int_a^b \frac{dQ}{T} = S_b - S_a$$

Clausius explained the nomenclature thus:

[2] This is Feynman's "property of the real world" alluded to earlier.

> "If we wish to designate S by a proper name we can say of it that it is the *transformation content* of the body, in the same way that we say of the quantity U that it is the *heat and work content* of the body. However, since I think it is better to take the names of such quantities as these, which are important for science, from the ancient languages, so that they can be introduced without change into all the modern languages, I proposed to name the magnitude S the *entropy* of the body, from the Greek word $\eta\tau\rho o\pi\eta$ a transformation. I have intentionally formed the word *entropy* so as to be as similar as possible to the word *energy*, since both these quantities, which are to be known by these names, are so nearly related to each other in their physical significance that a certain similarity in their names seemed to me advantageous" [22].

It is critical to realise that nothing at all is said about the absolute value of entropy; only the difference in entropy is defined. To understand the nature and full significance of entropy, it is necessary to consider, not just entropy changes in a particular system under examination, but all entropic changes in the universe due to thermodynamic action by the system. Any reversible process in a system in contact with a reservoir will cause an internal change in entropy of say $dS_{system} = + dQ_R/T$ where dQ_R heat is absorbed at temperature T. Since the same amount of heat is transferred from the reservoir the change in entropy of the reservoir is $dS_{reservoir} = - dQ_R/T$. Thus the nett change in entropy caused by the process for the whole universe is $dS_{system} + dS_{reservoir} = 0$. The change in entropy of the universe for a reversible process is zero. However, reversible processes are merely idealisations. All real processes are irreversible and the nett universal entropy change for irreversible processes is not zero. Clausius showed that for irreversible cycles the integral of the ratio of heat absorbed by the system to the temperature at which the heat is received is always less than zero:

$$_{I}\int \frac{dQ}{T} < 0$$

From this result it can be shown that for irreversible processes, $\mathrm{d}S_{system} + \mathrm{d}S_{reservoir} > 0$. Combining this with the above statement for reversible systems, we arrive a statement of what is known as the *entropy principle* and applies to all systems:

$$\Delta S_{universe} \geq 0$$

The upshot of this is that *any* process will at best cause no increase in the entropy of the universe, but all real processes will contribute to the increase of the entropy of the universe. This was realised by Clausius who presented his version of the first and Second Laws:

> "1. The energy of the universe is constant
> 2. The entropy of the universe tends toward a maximum" [22]).

Planck defined the entropy principle as:

> "Every physical or chemical process in nature takes place in such a way as to increase the sum of the entropies of all bodies taking part in the process. In the limit, i.e. for all reversible processes, the sum of all entropies remains unchanged" [63].

As an historical aside it is perhaps interesting to reflect on whether Carnot held some conception of what we now know as entropy. Carnot's theory of heat was primitive by modern standards. He considered that work done by a heat engine was generated by the movement of *calorique* from a hot body to a cooler body and was conserved in the transition. Clausius and William Thomson (Lord Kelvin) showed that the 'heat' in fact was not conserved in these processes. However, as Zemansky and Dittman observe:

> "Carnot used *chaleur* when referring to heat in general, but when referring to the motive power of fire that is brought about when heat enters an engine at high temperature and leaves at a low temperature, he used the expression *chute de calorique*, never *chute de chaleur*. It is of the opinion of some scientists that Carnot had at the back of his mind the concept of entropy, for which he had reserved the term *calorique*. This seems incredible, and yet is a remarkable circumstance that if the expression *chute de calorique* is translated as "fall of entropy," many of the objections to Carnot's work raised by Kelvin, Clapeyron, Clausius, and others are no longer valid" [88].

This is sustained when one considers that in Carnot's time the caloric theory of heat as a fluid dominated and much of Carnot's heat cycle theories were generated as analogies to water-wheel engines. What Carnot was trying to capture was a measure of heat quality that corresponded to the potential energy of water: the higher a stream feed enters above the base pool, the more work it can do per mass unit. This certainly corresponds to Lord Kelvin's "grade" of energy – that energy at a higher temperature, in some sense, has a higher quality.

In returning to the main discussion, we observe that we have arrived at the point where we can say that entropy is a thermodynamic property, with the particular characteristics described by the entropy principle. It is defined only by its changing. It is related to heat flows in a system and, like work and temperature, it is a purely macroscopic property governed by system state coordinates. But what are the microphysical actions that give rise to such macrocharacteristics? How are we to understand the underlying principles of entropy?

By stepping away from heat engines for a moment and examining what occurs at the microscopic level in examples of entropy increase in natural systems, correlations between entropy and order relationships are revealed. Consider the isothermal sublimation of dry ice to gaseous CO_2 at atmospheric pressure at 194.8K (−78.4°C). Heat is taken from the environment at this temperature increasing the internal energy of the solid to the point where molecules escape to become free gas. The enthalpy of sublimation[3] is 26.1 kJ/mol [77] which means the entropy increase associated with the sublimation of one gram of CO_2 can be calculated to be 3.04 J/K.

When forming a microphysical conception of entropy in such transitions, there is a tendency to associate increasing entropy with increasing disorder. Melting and sublimation are often used as illustrations (see [88]). However this approach can be somewhat misleading. Certainly phase transitions in melting a regular crystal to random liquid are associated with entropy increase, as is the transition of a material from a ferromagnetic to a paramagnetic state. These are examples of changes in microstructure from regularity to irregularity. But it is not the erosion of patterned regularity that directly accounts for entropy increase in these examples. Rather they are specific cases of a more general principle: that of increasing *degrees of freedom*.

In the sublimation example, the solid carbon dioxide leading up to and at the point of sublimation is a molecular solid. While forming regular structures at these lower temperatures the molecules are held together by very weak intermolecular forces (not by ionic or covalent bounds like metals or ice) and their dipole moments are zero. The molecules are held together in a solid state by induced dipole – induced dipole interaction[4] where instantaneous fluctuations in the electron density distribution in the non-polar compound produces a constantly changing bonding microstate. The induced dipole – induced dipole interaction is a very weak bond (0.4–4kJ/mol compared with

[3] The heat required for sublimation to occur.

[4] Also called London forces.

100–1000 kJ/mol for ionic or covalent bonds) that reduces the freedom of movement of the molecules. With strong bonds the freedom of movement is reduced by a greater degree. It is important to recognize that bonding does not primarily create regularities; it reduces the degrees of freedom. The order that emerges is wholly due to a restriction of kinetic phase space.

The tendency to equate order with regularity should be resisted. It is certainly the case that a highly regular system will possess fewer degrees of freedom than an irregular one. But regularity is not the only form of order. To further illustrate the contrasting notions of regularity and freedom let us return to the realm of heat engines and consider the very rapid, quasi-isothermal expansion of an ideal gas against a vacuum. In this case the result is similar to a slow isothermal expansion with the state variable entropy increasing as the volume in which the molecules are free to move increases. Here $\Delta S = \Delta Q/T$. It seems counterintuitive to say that there is more disorder in the final state than in the initial compressed state. There is the same number of molecules, with the same total kinetic energy moving in a random manner in both the initial and final states. There has been no change of structural regularity; only the volume has increased to provide greater freedom of movement to the molecules. When physical constraints are released, greater freedom is given to the microdynamics. This may also, in some systems, be reflected in an increase in disorder but it is the increased freedom that appears to be strongly correlated with an increase in the macroproperty entropy rather than some quantity *order*.[5]

We have reached the end of our introduction to entropy and the Second Law of Thermodynamics and we pause the list the important concepts to take from this section regarding entropy as a thermodynamic, macroscopic phenomenon. They are threefold. The first is that entropy and work are related concepts. Entropy limits the amount of work one may obtain from a physical system. The second is that for all real systems, the sum of the entropies of all bodies taking part in the system increases over time. This is the 'entropy principle'. Finally there exists a relationship between the macroscopic property of entropy and the degrees of freedom possessed by constituent microstates. We will look at this relationship between entropy and microstates in the next section; however before doing so, it is necessary, for completeness, to look at the Third Law of Thermodynamics.

[5] This is to say that greater concentration is not more orderly in any intuitive sense. (Consider millions grains of salt contained in a salt shaker and the same grains scattered on the table when the container is magically removed.)

Like the first two laws of thermodynamics, the third is a postulate and it relates to the absolute value of entropy. As noted above thermodynamic entropy only has physical significance when differences are considered. This is due to the integration operator[6]; the nature of absolute entropy is not defined. In 1906 Walther Nernst proposed a theorem to address the problem of determining an absolute value of entropy. "The gist of the theorem is contained in the statement that, as the temperature diminishes indefinitely the entropy of a chemically homogenous body of finite density approaches indefinitely near to a definite value, which is independent of the pressure, the state of aggregation and of the special chemical modification" [63]. The 'definite value' that entropy approaches is shown to be zero at absolute zero (0 K). Thus for homogenous solids and liquids (e.g. crystals) the theorem may be restated as*: Entropy approaches zero as temperature approaches absolute zero.* This is the third law of thermodynamics. We will examine the third law in more detail in Section 4.6.2.

Statistical Mechanics

The discussion at the end of the previous section concerning degrees of freedom and microphysical aspects of entropy was informal and qualitative in nature. In this section these considerations are extended and developed in a historical review of the formal relationship between entropy and system microstates. This review will prove valuable later when we consider the combinometric relationship between entropy and information.

The First Formulation

The first formulation of the relationship between thermodynamics of a system and its underlying molecular states was proposed by James Clerk Maxwell and Ludwig Boltzmann, though research into the underlying atomic kinetics of gases had commenced even earlier than Carnot's initial work on the laws of thermodynamics. In 1738 Daniel Bernoulli developed a particulate model of a gas which, assuming uniform particle velocity, predicted the inverse pressure – volume relationship at constant temperature and described the relationship of the square of (uniform) particle velocity to temperature. And, although similar work was carried out by W. Herepath (who, importantly, identified heat with internal motion) and J. Waterston (calculated specific

[6] On integration without limits the equation will produce an arbitrary additive constant.

heat relationships based on kinetics) during the first half of the nineteenth century, it wasn't until August Karl Krönig published a paper in 1856 detailing results akin to those of his kinetic-theory predecessors that an interest in microstate theories was kindled generally. The most likely reason for the less-than-receptive attitude towards a particulate theory of gases before Krönig is the sway that the caloric theory of heat held on the scientific community during the early 1800s.

Clausius continued the work on kinetic theory in the late 1850s, by taking into account the effect of molecular collision and by expanding internal energy calculations to include rotational and vibrational components, though, as with Bernoulli, the assumption of uniform molecular velocities (the *gleichberechtigt* assumption) remained. Clausius' consideration of the effect of molecular collision proved a vital point for it enabled future researchers, in particular Maxwell, to conclude that the uniform velocity assumption was unsustainable. If all molecules initially possessed identical velocities, they would not continue so because interactions between them would distribute the energy over a range of different velocities.

Maxwell was instrumental in developing a clear concept of the dynamic patterns that groups of molecules form in monatomic gases. He realised that while velocities of individual molecules were continually changing due to collisions, the velocity profile of the population at equilibrium was static and could be described. By considering subpopulations in velocity ranges, Maxwell developed what would lead to the first probabilistic account of molecular kinetics. The result of these considerations yielded the number of monatomic molecules in a discrete velocity range v to $v + \Delta v$ can stated as follows:

$$\Delta n = Ae^{-B(\dot{x}^2+\dot{y}^2+\dot{z}^2)}\Delta\dot{x}\Delta\dot{y}\Delta\dot{z}$$

Where: $\dot{x}, \dot{y}, \dot{z}$ are the velocity components in Cartesian space and A and B are two constants determined by total number molecules, total mass and the total kinetic energy. The relationship is known as Maxwell's Law.[7,8] The next major step toward a microdynamic account of thermodynamics occurred in 1868 with Boltzmann developing the kinetic theory of gases by constructing a significant generalisation of Maxwell's distribution law. Boltzmann's theory, like Maxwell's, allowed for non-uniform molecular velocities but also extended the notion to allow molecular non-uniformities of other types, specifically those

[7] Or, more completely, *Maxwell's Distribution Law for Velocities*.

[8] Given the form of the equation, it appears that Maxwell may well have been influenced by Gauss's then recent work on error distribution.

that were spatially dependent (eg. field effects). Rather than considering subpopulations in discrete velocity ranges, Boltzmann considered discrete ranges of *state* that extended the model to include energies beyond just kinetic energy. Later commentators, the Ehrenfests, have described the model thus:

> "If $\Delta\tau$ denotes a very small range of variation in the state of a molecule – so characterized that the coordinates and velocities of all atoms are enclosed by suitable limits, ... then for the case of thermal equilibrium
>
> $$f \cdot \Delta\tau = \alpha e^{-\beta\varepsilon} \cdot \Delta\tau$$
>
> gives the number of those molecules whose states lie in the range of variation $\Delta\tau$. Here ε denotes the total energy the molecule has in this state (kinetic energy + external potential energy + internal potential energy) and α and β are two constants which are to be determined just as in the case of Maxwell's law" [32].

In the appropriate limit, Boltzmann's distribution reduces to Maxell's distribution, hence the equation is known as the Maxwell-Boltzmann distribution law. This equation gives the energy distribution of the molecular system and has equilibrium as a stationary solution.

In 1872, Boltzmann undertook the development of a theorem to show that *only* equilibrium is a stationary solution, that all distributions will approach the Maxwell-Boltzmann distribution. As noted previously, in all real, irreversible processes entropy increases. Boltzmann defined a function, H, which could be applied to any distribution.

> "Consider a distribution, which may be arbitrarily different from a Maxwell-Boltzmann distribution, and let us denote by $f \cdot \Delta\tau$ the number of those molecules whose state lies in the small range $\Delta\tau$ of the state variables. Then we define the H-function as
>
> $$H = \sum f \log f \cdot \Delta\tau$$
>
> Where the sum is to be taken over all the possible domains of $\Delta\tau$" [32].

Boltzmann demonstrated that the H-function decreases monotonically with time so that for a time series $t_1, t_2, t_3 \ldots t_n$, the corresponding system H values are $H_1 \geq H_2 \geq H_3 \ldots \geq H_n$. On quick inspection we see that this behaves just as the negative value of thermodynamic entropy would and thus consider H to be an analogue of thermodynamic negentropy.[9] This gives us our first expression of entropy in microdynamic terms.

As a corollary to the theorem, Boltzmann showed that all non-Maxwell-Boltzmann distributions will, given time, approach a Maxwell-Boltzmann distribution. Further, Boltzmann showed that this is unique: all non-Maxwell-Boltzmann distributions will approach *only* a Maxwell-Boltzmann distribution. When the Maxwell-Boltzmann distribution is attained the equalities in the above H progression hold.

Since the work of Clausius, there has been embedded in the kinetic theory a postulate which eventually became the focus of criticism of the theory. The *Stosszahlansatz*[10] is an important assumption concerning the intermolecular collisions in a gas. In essence the assumption assigns equal probability to collisions. The number of collisions between two groups of molecules (e.g. those of two different velocities) is assumed to be independent of all factors except the relative densities of the two groups, total number of molecules and the proportional area swept out[11] by one of the groups of molecules. The inclusion of the *Stosszahlansatz* in Maxwell and Boltzmann's work led to a distribution that is stationary.[12] Questions soon arose regarding the capacity of a theory based on reversible kinetics to explain irreversible thermodynamic processes. How could a theory of stationary distributions deal with non-stationary processes, that is, processes with temporal direction?

However the H-theorem does not answer these questions founded in irreversibility arguments and two new objections demand consideration. The first was proposed by Josef Loschmidt in 1876. Termed *Umkehreinwand*[13], the objection was based on the reversible kinetics of the microstates. Consider the microstates of a gas that has reached equilibrium, that is at time n after the H-progression $H_1 \geq H_2 \geq H_3 \ldots \geq H_{n-1} = H_n$. Now consider an identical copy of this equilibrium gas but with all velocity vectors reversed. All molecules have the same speed as the original but the opposite direction. Because the H-theorem deals solely with scalar quantities, the H-function of the copy, H'_i, has the same value as the original H_i and since the mechanics of the system dictate energy conservation, the copy will therefore progress through

9 *Negentropy*, the negative value of entropy, will be discussed in Section 2.2.1.

10 Literally: Collision Number Assumption.

11 The volume "swept out" by a molecule can be considered to be all those points which lie in a path that is a collision-course for that molecule.

12 A stationary distribution is one that does not statistically change over time.

13 Reversibility Objection

the following phases: $H'_n = H'_{n-1} \cdots \leq H'_3 \leq H'_2 \leq H'_1$. So here we have an apparent instance of a gas at equilibrium spontaneously moving monotonically away from equilibrium; moving from a Maxwell-Boltzmann distribution to a non-Maxwell-Boltzmann distribution.

The second objection, the *Wiederkehreinwand*[14], proposed by Ernst Zermelo in 1896, attacks the problem from a different angle. Henri Poincaré showed in 1889 that for energetically conservative systems bound in a finite space, the trajectory in phase space of a system starting from a specified initial state will, except for a 'vanishingly small' number of initial states, eventually return arbitrarily close to the initial state. This is known as Poincaré's Recurrence theorem. Zermelo's argument employed Poincaré's Recurrence theorem to point out that if we take a gas system that is not at equilibrium, a non-Maxwell-Boltzmann distribution, then at some state in its future the state of the system will be arbitrarily close to its initial state. This is at odds with Boltzmann's claim that all non-Maxwell-Boltzmann distributions move monotonically to a Maxwell-Boltzmann distribution and stay there because they are at equilibrium.

These objections led Boltzmann to a revised, probabilistic formulation of kinetic theory.

The New Formulation

In 1877 Boltzmann issued a reply to Loschmidt's *Umkehreinwand.* Boltzmann argued that, while it is true that the evolution of a system from a specific initial microstate does depend on exactly those initial conditions, it is possible to provide a general account of all gases by adopting a statistical approach. Every individual microstate has the same probability of occurrence, but the microstates that correspond to the macroequilibrium conditions are more numerous than those that correspond to non-equilibrium macrostates at any given time instance. That is, for a number of arbitrarily chosen initial microstates, many more initial microstates corresponding to non-equilibrium macrostates will tend to microstates corresponding to equilibrium macrostates than vice versa.

Boltzmann formulated a model based on dividing microstate space into small, discrete ranges: spatial momentum ranges. The project then became to work out, given macro constraints (total energy, total number of molecules), how many ways can the molecules be distributed across these ranges? A distribution is the number of particles in each

[14] Recurrence Objection

range. A number of distinct system states can have the same distribution, simply by swapping distinct particles between particle states. Boltzmann demonstrated that, if the probability of a distribution is defined by the number of ways a distribution can be constructed by assigning molecules to ranges then there exists a most probable distribution and that in the limit of the number of ranges going to infinity and range size going to zero this distribution uniquely corresponds to the Maxwell-Boltzmann distribution.

Boltzmann defined W, the probability that a system is in a particular microstate using the distribution definition.[15] Combining this with the H-theorem and the notion of thermodynamic entropy, he arrived at the following kinetic description of thermodynamic entropy:

$$S = -K \log W$$

The term W can be calculated as follows:

$$W = \frac{N!}{\prod_i N_i!}$$

where N is the total number of systems and N_i is the number of systems in a particular microstate i. This new formulation did not, however, stop criticism based on *Umkehreinwand*-like reversibility arguments. As Sklar observes,

> "Boltzmann's new statistical interpretation of the H-theorem seems to tell us that we ought to consider transitions from microstates corresponding to a non-equilibrium macrocondition to microstates corresponding to a condition closer to equilibrium as more 'probable' than transitions of the reverse kind. But if, as Boltzmann would have us believe, all microstates have equal probability, this seems impossible. For given any pair of microstates, S_1, S_2 such that S_1 evolves to S_2 after a certain time interval, there will be a pair S_1', S_2' – the states obtained by reversing the directions of motion in the respective original microstates while keeping speeds and positions constant – such that S_2' is closer to equilibrium than S_1' and yet S_2' evolves to S_1' over the same time interval. So these 'anti-kinetic' transitions should be as probable as 'kinetic' transitions" [73].

---

[15] The number of ways a system can be in a particular state divided by the total system permutations.

Eventually Boltzmann gave up notions of monotonic evolution of non-equilibrium systems toward a Maxwell-Boltzmann distribution. Instead he considered that over large amounts of time, systems would remain close to equilibrium for most of the time, occasionally drifting away from equilibrium distribution an arbitrary distance and returning at a frequency that is inversely proportional to the distance away from equilibrium. This, Boltzmann argued is consistent with Poincaré's Recurrence theorem.

With respect to our analysis of information theory, the crucial outcome of Maxwell and Boltzmann's work as described in this section is the construction of a formal quantifier of the number of unique distributions that distinct system states may have. The development of a notion of Boltzmann's thermodynamic probability, W, provides us with a means of counting distinct macrostates and, as we shall see later, it is a system's capacity to exist in uniquely identifiable states that governs the quality of information it is capable of possessing.

We also see that the notion of distinguishability is crucial. Indeed it will form a fundamental part of my account of information (see Section 3.1). For the ability to distinguish between particles in different energy ranges potentially allows one to extract work by applying this information to a sorting process. However this threatens to violate the Second Law. This 'paradox' is known as Maxwell's Demon, but we will see that, instead of being a paradox, it is instead a demonstration that information can do physical work in a system.

Maxwell's Demon

At the conceptual centre of thermodynamic considerations of information is the relationship between entropy and information. Historical consideration of the nexus arose as the result of a thought experiment proposed by Maxwell in 1871 in his *Theory of Heat.* Maxwell considered a gaseous system contained at equilibrium in an insulated vessel consisting of two chambers, A and B, separated by a trap door. Stationed at the trap door was a *Demon*: a being "whose faculties are so sharpened that he can follow every molecule in its course" (Maxwell quoted in [51]). Such a Demon would operate the trap door (without friction or inertia) permitting only faster molecules to travel from A to B and slower molecules to travel from B to A.[16] In Fig. 2.3. below, a schematic representation of the system is shown with the varying molecular velocities represented by varying arrow sizes.

[16] "Fast" and "Slow" could be designated as being greater-than and less-than the system average molecular velocity respectively.

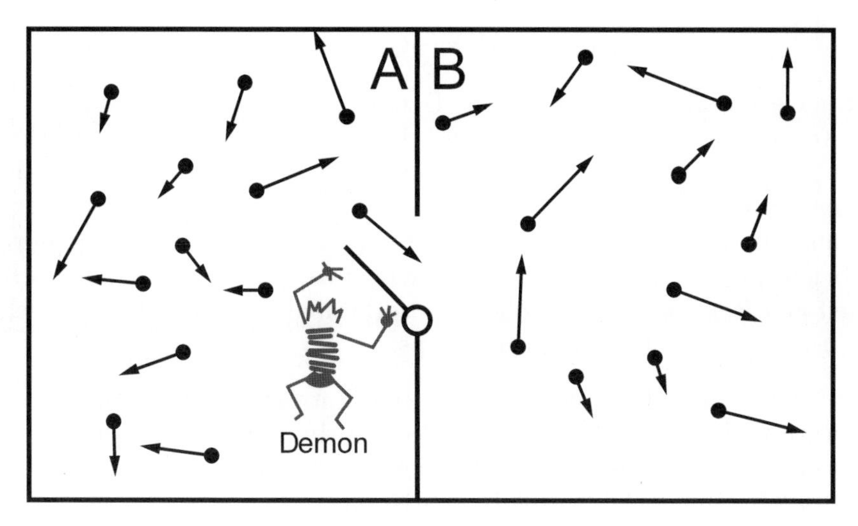

Fig. 2.3. Maxwell's Demon

The conceptual paradox rests in the fact that as time progresses, increasingly more fast molecules will occupy chamber B and the slower molecules will occupy chamber A. If the total population is sufficiently large and the initial velocity distribution was symmetric, approximately equal numbers of molecules will eventually occupy both chambers but the molecules in chamber B will have a greater total kinetic energy than those in chamber A resulting in increased temperature. This is in conflict with Clausius' form of the Second Law of Thermodynamics in that it is equivalent to heat flow from a low temperature to a high temperature with no other effect.

The partitioning of energy by the Demon could also manifest as an increase in pressure that could be used to do work – pushing a piston for example. Thus this formulation is in direct contradiction to Planck's interpretation of the Second Law, for if we reset the piston, allow the molecules to remix and start all over again, we would have a perpetual motion machine of the second kind.[17] If this process is performed isothermally at temperature T (that is in contact with an infinite reservoir at temperature T) and produces work W with no waste heat, then the heat transferred from the reservoir to the gas is Q=W which satisfies the first law. However, the change in entropy is

[17] A distinction between types of perpetual motion machine was introduced by W. Ostwald late in the 19th century. A *perpetuum mobile* of the first kind is one that violates the first law of thermodynamics. A *perpetuum mobile* of the second kind violates the Second Law.

$$\Delta S = -Q/T.$$

It is clear that the actions of the Demon constitute a sorting process; faster molecules are separated from slower ones and vice versa, so that after some time they are divided into two groups. To maintain the integrity of the Second Law, entropy must somewhere be produced in a quantity at least as great as that reduced by the sorting. The most obvious place to look for this increase in entropy is in the Demon itself. Leff and Rex consider the following isothermal 'pressure-Demon' cycle consisting of the following three steps:

> "(a) The Demon reduces the gas entropy at fixed temperature and energy by letting molecules through the partition in one direction only. This sorting process generates pressure and density differences across the partition.
>
> (b) The gas returns to its initial state by doing isothermal work on an external load. Specifically; the partition becomes a frictionless piston coupled to a load, moving slowly to a position of mechanical equilibrium (away from the container's centre) with zero pressure and density gradients across the piston. The piston is then withdrawn and reinserted at the container's centre.
>
> (c) The Demon is returned to its initial state" [51].

Thermodynamic analysis of the cycle reveals that, if we are to preserve the integrity of the Second Law, the entropy of the Demon must increase in order to 'pay' for the entropy reduction of the gas in step (a). The work done in (b) is compensated for by heat transfer $Q = W$ from the reservoir. There is no change in the load's entropy. If the Demon is to continue its sorting function through repeated iterations of the cycle, the entropy that it accrues in step (a) must be reduced by a resetting process otherwise the accumulation of entropy would eventually render it inoperable. Hence the resetting of the Demon in step (c), which must also be a thermodynamic process. So we can assume that the Demon returned "to its initial state by energy exchanges with the reservoir and a reversible work source, with work E being done on the Demon. The Demon's entropy decrease here must be compensated for by an entropy increase in the reservoir. We conclude that resetting the Demon results in heat transfer to the reservoir"(ibid). Leff and Rex continue,

> "Overall, in (a)–(c) the entropy change of the universe equals that of the reservoir. The Second Law guarantees this is non-negative; i.e., the reservoir cannot lose energy. The cyclic process

results in an increased load energy and a reservoir internal energy that is no lower than its initial value. The first law implies that the work source loses sufficient internal energy to generate the above gains; in particular, the source does positive work in (c). The relevant energy transfers during the cycle are: Work $W > 0$ by gas on load, work $E > 0$ by work source on Demon, and energy $E - W \geq 0$ added to the reservoir. The entropy change of the universe is $(E - W)/T \geq 0$, where T is the reservoir temperature"(ibid).

We see that in Leff and Rex's cycle, if the Second Law is preserved, the resetting of the Demon is of fundamental importance. Such considerations of the importance of resetting or erasure also figure centrally in work by recent researchers constructing computational Demon models. Landauer introduced the concept of "logical irreversibility": the transformation of any computational memory state to an erased one is a many-to-one mapping which has no unique inverse. Similarly, Bennett showed that, in its simplest form, the Demon's memory may be considered to be a two-state system: 'did nothing'/ 'let through'. Prior to making a measurement the Demon is constrained to be in just one state: the reference or 'did nothing' state. On measuring a molecule, the Demon has the dimensionality of its state space doubled so that it may now be in either one or the other state. Thus Bennett takes erasure to be the recompression of state space to the reference state, regardless of the prior measurement states. This compression is logically irreversible and generates an entropy increase in the reservoir.

Some researchers[18] questioned the possibility of a Demon operating as required by the thought experiment since, located inside the gas, it must be continually bombarded by gas molecules and absorbing energy. This bombardment would interfere with the Demon making accurate measurements. Others pointed out the need for a means of measuring molecular velocities and the need for some kind of memory faculty. In particular, Leo Szilard demonstrated, using a simplified one molecule model, that the process of measuring the position and velocity of the molecule generated at least as much entropy as was reduced in the gas. Szilard's model [79] provides a tractable simplification of Maxwell's Demon embedded in a work cycle that enables us to see the relationship between information, measurement and the thermodynamics of the Demon.

Imagine a vertical cylinder that can be horizontally divided into two, not necessarily equal, sections with volumes V_1 and V_2 by the

[18] Smoluchowski and Feynman

insertion of partition. The cylinder, which is in contact with an infinite reservoir at temperature T, contains a single molecule which is free to move about the cylinder under thermal motion. On the insertion of the partition, an observer notes whether the molecule is caught in the upper or lower sections of the cylinder. This step is *measurement* and important to Szilard's preservation of the Second Law. The partition is now free to move the level of the cylinder, acting as a piston, as the molecule-gas undergoes isothermal expansion. If the molecule is caught in the upper section, the piston will move slowly downward changing the volume from V_1 to $V_1 + V_2$.[19] A weight could be attached to the piston to produce work. On completion of the expansion the partition is removed and the process is repeated ad infinitum with the weight being attached in a manner that will ensure that it is always displaced upwards. This attachment will require a binary switch that will be set by the observer depending on the direction of motion of the piston (i.e. whether the molecule is in the upper or lower part of the cylinder).

Without more explanation than "reasonable assumption", Szilard compensates the decrease in system entropy with the entropy increase generated by the measurement process, saying,

> "One may reasonably assume that a measurement procedure is fundamentally associated with a certain definite average entropy production, and that this restores concordance with the Second Law. The amount of entropy generated by the measurement may, of course, always be greater than this fundamental amount, but not smaller" [79].

In the binary-state monomolecular system, Szilard calculated this entropy generated by measurement to be at least equal to k log 2 (where k is a constant). Memory was also an important component of Szilard's model. If we denote the physical position of the molecule by independent variable x and a dependent measuring variable by y, then when x and y are coupled (measurement), x sets the value of y. The variables are then uncoupled and x can vary while y keeps the value it had at coupling. This is a form of memory and it is crucial for the cycle if it is to produce work. So although Szilard does not explicitly connect information with entropy, his analysis of measurement, utilisation of measurement and memory certainly implies the existence of a role that most would intuitively think of information as filling.

Twenty-one years later, Leon Brillouin directly examined the relationship between information and entropy. Brillouin expanded Szilard's

[19] It should be noted here that Szilard ignores the effects of gravity.

work on the entropy of measurement by considering the information gain associated with measurement. Following Shannon (see Section 2.2.2), Brillouin, from the outset, derives a definition of information based on statistical considerations:

> "Let us consider a situation in which P_0 different possible things might happen, but with the condition that these P_0 possible outcomes are equally probable a priori. This is the initial situation, when we have no special information about the system under consideration. If we obtain more information about the problem, we may be able to specify that only one out of the P_0 outcomes be actually realized. The greater the uncertainty in the initial problem is, the greater P_0 will be, and the larger will be the amount of information required to make the selection. Summarizing, we have:
>
> Initial situation: $I_0 = 0$ with P_0 equally probable outcomes;
>
> Final situation: $I_1 \neq 0$ with $P_1 = 1$, i.e. one single outcome selected.
>
> The symbol I denotes information, and the definition of information is
>
> $$I_1 = K \ln P_0$$
>
> Where K is a constant and 'ln' means the natural logarithm to the base e" [11].

The relationship that information has with entropy, according to Brillouin, is that of a reversal of sign.[20] He attributes Szilard with showing that Maxwell's Demon "actually transforms 'information' into 'negative entropy"' [10]. By constructing a model in which the Demon has a single photon source (a high filament temperature electric torch) to identify molecules, Brillouin shows that the torch generates negative entropy in the system. The Demon obtains "informations" concerning the incoming molecules from this negative entropy and acts on these by operating the trap door. The sorting rebuilds the negative entropy, thus forming a cycle:

$$\text{negentropy} \rightarrow \text{information} \rightarrow \text{negentropy}$$

The notion of negentropy "corresponds to 'grade' of energy in Kelvin's discussion of the 'degradation of energy"'(ibid).

Brillouin undertakes an entropy balance on the Demon system to quantify the negentropy transformations. The torch is a radiation

[20] Positive or negative.

source not at equilibrium and so "pours negative entropy into the system". If the filament is at a temperature T_1 and radiates energy E, then the radiation emission is accompanied by an entropy increase $S_f = E/T_1$. As just noted, since $T_1 \gg T_0$ (the system temperature) the filament is a source of relatively high negative entropy radiation. If the Demon does not act, the energy E is dissipated with a global entropy increase of $S = E/T_0 > S_f > 0$. However, if the Demon is to act, the minimum requirement for the Demon to determine the state of an approaching molecule it that at least one quantum of energy be scattered by the molecule and be absorbed by the Demon's eye. For the light to be distinguishable from the background black body radiation, the energy of the photon, hv_1, must be much greater than background, kT_0, where h and k are Planck's and Boltzmann's constants respectively. Thus the entropy increase of the Demon will be $\Delta S_d = hv_1/T_0 = kb$ where b is the ratio of photon energy to background radiation energy $(hv_1/kT_0 \gg 1)$.

Once the Demon has information concerning the molecule, it can be used to reduce system entropy: information is converted to negentropy. On receipt of the information, the state of the system is more completely specified, hence the number of possible molecule arrangements, "complexions", has been reduced. Let P_0 represent the initial total number of microstate configurations (equivalent to Boltzmann's thermodynamic probability W) and P_1 be the number of microstate configurations after the receipt of information. Thus we can define p to be the reduction on the number of complexions: $P_0 - P_1$. By Boltzmann's formula, $S_0 = k \ln P_0$ and $S_1 = k \ln P_1$. Thus the change in entropy on sorting becomes:

$$\Delta S_i = S_1 - S_0 = k \ln(P_1/P_0) \approx -k(p/P_0) < 0,$$

(since for most cases $p << P_0$). Calculating the total entropy balance we have:

$$\Delta S_d + \Delta S_i = k(b - p/P_0) > 0,$$

since $b \gg 1$ and $p/P_0 \gg 1$. Brillouin says,

> "The final result is still an increase of entropy of the isolated system, as required by the second principle. All the Demon can do is recuperate a small part of the entropy and use the information to decrease the degradation of energy.
>
> In the first part of the process ..., we have an increase of entropy ΔS_d, hence, a change ΔN_d in the negentropy:
>
> $$\Delta N_d = -kb < 0, \text{ a decrease.}$$

From this lost negentropy, a certain amount is changed into information, and in the last step of the process ... this information is turned into negentropy again:

$$\Delta N_i = k(p/P_0) > 0, \text{ an increase" [10].}$$

This is the justification for the negentropy/information cycle stated earlier.

Cursory study of Brillouin's model reveals a conspicuous detail. Brillouin talks of the "information turned into negentropy", "negentropy changed into information" and the "transformation of information into negentropy". Nowhere does Brillouin equate information and negentropy as Schrödinger does (see next section). In the measurement step of Brillouin's model, only "a certain amount" of the "lost negentropy" is changed into information. Perhaps Brillouin intends that the remainder accounts for the information in the Demon, assuming it is physical. However, if this is the case, he does not state this explicitly.

As noted previously, Brillouin defines information in a Shannon-like manner as the logarithm of the number of equal a priori possibilities. Further, he distinguishes between two classes of information:

"1. Free information I_f, which occurs when the possible cases are regarded as abstract and have no physical significance.

2. Bound information I_b, which occurs when the possible cases can be interpreted as complexions of a physical system. Bound information is thus a special case of free information" [11].

He makes this distinction in order to draw a connection between thermodynamic entropy and information. This is an attempt to avoid thorny epistemological issues concerning information, such as the intractability of the determining the information gain when a person hears some news or the information loss when someone forgets. Only bound information is associated with entropy changes of a system. Consider a system in which the "complexions" (P_0 and P_1) of two temporal states of the system (corresponding to times t_0 and t_1) are equally probable cases. Then if $P_1 < P_0$ the physical entropy of the system will decrease and "the entropy decrease when information is obtained, reducing the number of complexions, and this information must be furnished by some external agent whose entropy will increase. The relation between the decrease in entropy of our system and the required information is obvious, for

$$I_{b1} = k(\ln P_0 - \ln P_1) = S_0 - S_1,$$

or

$$S_1 = S_0 - I_{b1}.$$

the bound information appears as a negative term in the total entropy of the physical system, and we conclude:

Bound information = decrease in entropy S = increase in negentropy N,

where we define negentropy to be the negative of entropy" (ibid).

Thus, on this account not all information is negentropy, only bound information. Brillouin calls the relationship between bound information and entropy the "negentropy principle of information". However has we have seen above in the conversion cycle in Maxwell's Demon,

$$\text{negentropy} \rightarrow \text{information} \rightarrow \text{negentropy},$$

the relationship is not truly one of identity; it is not even conservative. This leads me to judge that Brillouin negative entropy principle does not provide a truly foundational account of the nature of information.

Brillouin's version of the principle the negentropy principle of information is akin to some later work of Erwin Schrödinger's in which Schrödinger examines the somewhat stronger relationship between order and negative entropy. This work is examined in the following section.

Schrödinger

Additional thoughts on the physical nature negentropy come from Erwin Schrödinger. In his 1944 book *What is Life?* Schrödinger considered the relationship between entropy and order. In trying to work towards an answer to the question posed in the title of his book, Schrödinger observed that living matter was ordered in a way that evaded the 'decay to equilibrium'. He says, "Life seems to be orderly and lawful behaviour of matter, not based exclusively on its tendency to go over from order to disorder, but based partly on existing order that is kept up" [70]. Schrödinger notes that the systems that tend towards equilibrium, move towards a state of maximum entropy, which, he notes, is a state of death. Living systems maintain ordered integrity not just by energetic intake, but by drawing from the environment negative entropy, thus staving off the tendency to maximum entropy. Schrödinger equates negentropy with order by considering Boltzmann's equation to be broadly interpretable as entropy $= k\log(D)$, where D represents disorder. He then notes the following:

> "If D is a measure of disorder, its reciprocal, $1/D$, can be regarded as a direct measure of order. Since the logarithm of $1/D$ is just minus the logarithm of D, we can write Boltzmann's equation thus:
>
> $$-\text{entropy} = k \log(1/D).$$
>
> Hence the awkward expression 'negative entropy' can be replaced by a better one: entropy, taken with the negative sign, is itself a measure of order"(ibid, p.73).

What Schrödinger adds to the thermodynamic/Statistical Mechanics approach to information theory is a direct identification of the negative sign of entropy with order. The extension of this relationship to information relies on the nature of the correlation of information and order. I do not equate the two (and nor, I feel, does Schrödinger). But I do believe that they are related via the notion of degrees of freedom as outlined previously. This will be discussed further in Section 4.6.

Signpost

In this book I am attempting to construct a theory of the physical foundations of information. This section has been an historical examination of the study of the relationship between thermodynamic entropy and information. Early in the section we looked at the development of thermodynamics with a special interest in the Second Law and established the relationship between entropy and work and between entropy and order, or more accurately, between entropy and degrees of freedom. Then we reviewed the discovery of the relationship between the macroproperty entropy and a microsystem's states noting the important role that combinometrics plays relating entropy to the microdynamics of a system. Discussion of the role of measurement in simple Maxwell's Demon systems led us to consider the application of information regarding the microstate of a system to extract work in an apparent violation of the Second Law and, finally, to the relationship between negative entropy and information.

Four fundamental concepts should be taken from this section for use in the development of my theory. The first is the relationship between entropy and the number of identical states in a system as defined in Boltzmann's theorem. We will see in Section 4.5 that entropy is primarily about counting distinguishable possible states and that, due to the intimate relationship between entropy and information as noted by Brillouin and Schrödinger, information is also combinometrically grounded.

The second important concept introduced in this section is 'measurement'. Szilard, in considering Maxwell's Demon, noted that the measurement process is essential for the preservation of the Second Law in a Demon cycle. Measurement (and its limitations) is a significant element in the development of my account of information, though in somewhat different ways. The place of measurement plays a fundamental role in Szilard's formulation in that measurement generates at least as much entropy as was reduced in the gas thus preserving the Second Law. Under my interpretation, measurement bounds the amount of information that may be apprehended from an object. In Section 3.2 I will discuss the apprehension of information from an informatic object and in Section 4.5.2 I will further discuss measurement, Demons and information.

The third concept introduced in this section is 'memory'. Memory is closely tied to measurement for if one is to say that the value of a fluctuating parameter of a system, y, is different from or the same as an earlier measurement of the parameter, x, then some storage faculty is required. The concept of distinguishability of temporal states of a system is developed in Section 3.1, and the relationship between information and memory is discussed in Section 4.2.2

The final concept introduced here is the relationship between work and information. In Section 4.6 we will examine the effect increasing information on work capacity by looking at Gibbs' Paradox and through further consideration of Maxwell's Demon.

Boltzmann's H-theorem proved to be an inspiration to others who were thinking about information outside the field of thermodynamics with the concept of thermodynamic probability transplanted into realms where occurrence probabilities apply, for example, in the transmission of electrical communication signals. This field is widely referred to as "Information Theory" and is reviewed in the next section.

2.2.2 Information (Communication) Theory

Although much development of theories of information was undertaken by Leo Szilard and other writers[21] in the first half of the 20th Century, paternity of modern information theory is generally assigned to Claude Shannon. In his 1948 article "A Mathematical Theory of Communication", Shannon addresses the problem of communication: the exact or

[21] N. Weiner, H. Nyquist, R.V.L Hartley, J. von Neumann, etc.

approximate transmission and reception of a message from a generating source. It is from this perspective that he develops his theory of information.

Shannon's concerns are purely with engineering. Semantic aspects of information, e.g. the content of the message and its meaning to the recipient, are irrelevant to the problem; the theoretic description which he seeks must function not just for an actual, individual message but for each possible message that may be sent. The approach is fundamentally probabilistic. To develop his theory, Shannon examines the output of a discrete information source that generates a message as a Markov Process.[22] Each possible message that can be generated has associated with it a probability p_i of its occurrence and there are n such messages.

Shannon attempts to define a quality which will measure the amount of information generated by the process. He searches for a function $H(p_1, p_2, \dots p_n)$ that will quantify our reduction in uncertainty on receiving the message subject to the following desiderata:

1. H should be continuous in p_i.
2. If all the p_i are equal, the H should be a monotonic increasing function of n.[23]
3. Each event (symbol generation) should be capable of being linearly decomposed into two or more constituent events with their own proportional probabilities. [72].

Shannon concludes that the only function that satisfied the criteria was of the form:

$$H = -K \sum_{i=1}^{n} p_i \log p_i$$

where K is a constant according to units chosen (i.e. the base of logarithm used). In what is more than a nod to Boltzmann, Shannon assigns H to be the entropy of the set of probabilities $(p_1, p_2, \dots p_n)$. This function has several properties which Shannon believes further substantiates its use as a measure of information:

> 1. $H = 0$ iff all the p_i but one are zero, this one having value 1. That is, information is zero if the outcome is already certain.

[22] A Markov process is a random process whose future probabilities are determined by its most recent values.

[23] This means that as the number of equiprobable symbols increases, there is more uncertainty.

2. For a given n, H is maximum and equal to $\log n$ when all the p_i are equal, that is, $\frac{1}{n}$. This is the case of maximal uncertainty.
3. The uncertainty[24] of a joint event is less than or equal to the sum of the individual uncertainties, having equality only when the two events are independent.
4. Almost as a corollary to point 2, any change towards the equalization of the probabilities $p_1, p_2, \ldots\ p_n$ increases H.
5. The uncertainty of a joint event x, y is the uncertainty of x plus the uncertainty of y when x is known. Thus the uncertainty of y is never increased by knowledge of x [72].

Like Szilard and Brillouin, Shannon uses a Boltzmann-like entropy theorem to quantify information capacity. In Shannon's account information is evaluated by summing uniquely identifiable distributions. He also opens up the possibility of evaluating the mutual or conditional information in multiple messages by calculating joint entropies or chaining their entropies and Shannon's approach has proved to be valuable in practical applications in the fields of communications and electrical engineering.[25]

Shortly after Claude Shannon proposed his account of information researchers in mathematics and the nascent field of computer science began thinking about information in a manner totally different from the Boltzmann-based approaches of Szilard, Brillouin and Shannon. This third and markedly novel approach to the quantification of information emerged from the mid 1950's and on from the work of Kolmogorov, Solomonoff and Chaitin. Though still having its conceptual origins in probability theory, Algorithmic Information Theory provides, through applied computability theory, a method of measuring the intrinsic information in an object's description. The following section reviews this approach to providing an account of information.

2.2.3 Algorithmic Information Theory

Algorithmic Information Theory was born out of inconsistencies that arise between intuitive notions concerning regularity and answers provided by standard probability theory. In order to illustrate the nature of these dissatisfactions, consider a binary experiment (coin-toss) con-

[24] Shannon uses *uncertainty* and *entropy* interchangeably.

[25] Constraints on the maximum possible rate of transmission of information via a standard modem are determined by Shannon's theorem.

ducted 23 times. Now imagine we obtained the following results from three trials:

Trial 1: 1 0 1 1 0 0 0 1 1 1 1 0 0 1 1 0 0 0 1 1 0 1 0

Trial 2: 1 1 1 1 1 1 1 1 1 1 1 1 1 1 1 1 1 1 1 1 1 1 1

Trial 3: 0 0 1 1 0 1 1 1 0 0 1 0 1 1 1 0 1 1 1 1 0 0 0

Probability theory tells us the probability of obtaining the sequence shown in trial 1 is 2^{-23} (about 0.00000011920). This value is also true for trials 2 and 3. However, there is something intuitively unsettling about accepting that the first and second results are equally probable. The first result "looks" more random; it appears more consistent with the process that is supposed to have generated the string. If presented with the second sequence as a result of a coin-toss, an observer may be entitled to doubt the fairness of the coin. The third trial may at first glance appear random, but there is also regularity in this sequence. The first trial was generated by tossing a 20-cent coin. The second is obviously just a series of 23 ones. Listing the integers 0 to 8 in binary format created the third series.[26] The issue here is that neither of the probability values of trial 1 nor of trial 2 tells us anything about the inherent order that is present in each sequence, independently of how they were generated. The use of a probability method assuming equiprobable occurrences will not truly account for the information embedded in the order in the sequences. Something more is required, something that takes into account the generation process. Algorithm Information Theory represents an attempt to meet that need.

The theory was proposed separately by R. Solomonoff of the Zator Company in 1960, A.N Kolmogorov in 1965 and by G. Chaitin also in 1966 and thus is justifiably called by some the Solomonoff-Kolmogorov-Chaitin theory of information. However it is more common to refer to the entire field of Algorithmic Information Theory and descriptor complexity as "Kolmogorov Complexity". Once developed, I will show how the asymmetry, foundational account of information is also compatible with Algorithmic Information Theory.

Before examining Algorithmic Information Theory in detail it is necessary to first look at two underlying notions: Turing Machines and randomness.

---

[26] This is the beginning of Champernowne's Number in binary format. Champernowne's number base 10 is 0.123456789101112131415161718192021.

Turing Machines

The notion of a universal computing machine arose initially out of early work by Alan Turing in the consideration of Computable Numbers [81]. The model of a 'Turing Machine' is important to Algorithmic Information Theory because it provides a rigorous definition of computability by developing a mathematically well-defined means of generating descriptive integer sequences, or "strings". Turing's original conceptual engine consisted of an automated machine (a-machine) that mimicked human pen and paper implementation of an algorithm. It consisted of simple acts of iterated reading or writing of a symbol and the transference of 'focus' from one place on the paper to a different place on the paper, usually thought of as a continuous tape. The action of the machine depends solely on the current state of the a-machine and the symbol at the momentary focal location. Turing writes,

> "We may compare a man in the process of computing a real number to a machine which is only capable of a finite number of conditions $q_1, q_2, \ldots, q_R$ which will be called "m-configurations". The machine is supplied with a "tape" (the analogue of paper) running through it, and divided into sections (called "squares") each capable of bearing a "symbol". At any moment there is just one square, say the r-th, bearing the symbol (r) which is "in the machine". We may call this square the "scanned square". The symbol on the scanned square may be called the "scanned symbol". The "scanned symbol" is the only one of which the machine is, so to speak, "directly aware". However, by altering its m-configuration the machine can effectively remember some of the symbols which it has "seen" (scanned) previously. The possible behaviour of the machine at any moment is determined by the m-configuration q_n and the scanned symbol (r). This pair q_n, (r) will be called the "configuration": thus the configuration determines the possible behaviour of the machine. In some of the configurations in which the scanned square is blank (i.e. bears no symbol) the machine writes down a new symbol on the scanned square: in other configurations it erases the scanned symbol. The machine may also change the square which is being scanned, but only by shifting it one place to right or left. In addition to any of these operations the m-configuration may be changed. Some of the symbols written down will form the sequence of figures which is the decimal of the real number which is being computed. The others are just

rough notes to "assist the memory". It will only be these rough notes which will be liable to erasure" [81].

Many variants of Turing Machines have since been proposed, principally involving differing numbers of tapes and state representations. These variations do not affect the underlying principle of computation; their chief advantage lies with explanatory powers. Working memory, the "rough notes" that Turing refers to, is often included on a separate tape and the written output is presented separately. These features are included in a variant proposed by Chaitin,

> "Each Turing machine has three tapes: a program tape, a work tape, and an output tape. There is a scanning head on each of the three tapes. The program tape is read-only and each of its squares contains a 0 or a 1. It may be shifted in only one direction. The work tape may be shifted in either direction and may be read and erased, and each of its squares contains a blank, a 0, or a 1. The work tape is initially blank. The output tape may be shifted in only one direction. Its squares are initially blank, and may have a 0, a 1, or a comma written on them, and cannot be rewritten. Each Turing machine of this type has a finite number n of states, and is defined by an nx3 table, which gives the action to be performed and the next state as a function of the current state and the contents of the square of the work tape that is currently being scanned. The first state in this table is by convention the initial state. There are eleven possible actions: halt, shift work tape left/right, write blank/0/1 on work tape, read square of program tape currently being scanned and copy onto square of work tape currently being scanned and then shift program tape, write 0/1/comma on output tape and then shift output tape, and consult oracle. The oracle is included for the purpose of defining relative concepts. It enables the Turing machine to choose between two possible state transitions, depending on whether or not the binary string currently being scanned on the work tape is in a certain set, which for now we shall take to be the null set" [21].

The capacity for a Turing machine to algorithmically generate an output string in a rigorously defined manner makes it invaluable for developments in Algorithmic Information Theory. We will use the Chaitin defined machine in future references to Turing machines.

Distinguishing between a string that has been generated algorithmically by a Turing machine and one that is in some sense random is significant in Algorithmic Information Theory. This relies on the ability to measure, or at least detect, randomness. We examine this in the next section.

Randomness

Randomness lies at the heart of considerations of Algorithmic Information Theory; indeed, it is in the defining of "randomness" that the theory's core is founded. We commence by considering the occurrence of numerals and groups of numerals in sequences. We call these sequences *strings*. Smaller contained sections of these sequences are termed subsequences or substrings.

One simple test of randomness in a string expressing a number is to show that, at least statistically, it is a Normal Number.[27]. It is insufficient to simply require that all possible states in a sequence be equiprobable. Examples abound which pass this test and yet are clearly non-random; Champernowne's number is one example.

As we shall see later, an appreciation of the difference between a state-generated sequence and a string (or number) formed by a concatenation of symbols representing those states is crucial to understanding information as it is used in Algorithmic Information Theory.

Richard von Mises' interpretation of randomness was an important starting point for the development of Algorithmic information Theory. Von Mises was a mathematician who specialised for the most part in hydrodynamical and aerodynamical studies but is possibly best remembered for his continuing work on the frequency interpretation of probability commenced by Venn. Kolmogorov has the following to say regarding von Mises,

> "... the basis for the applicability of the results of the mathematical theory of probability to real 'random phenomena' must depend on some form of the frequency concept of probability, the unavoidable nature of which has been established by von Mises in a spirited manner" [47].

We will note in Section 2.3.2 that von Mises' definition of randomness requires that, for any attribute under consideration in a string,

[27] A Normal Number is an irrational number for which any finite pattern of numbers occurs with the expected limiting frequency in the expansion in a given base. For example, for a normal decimal number, each digit 0–9 would be expected to occur 1/10 of the time, each pair of digits 00–99 would be expected to occur 1/100 of the time, etc [86]

the limiting relative frequency of any subsequence be the same as the limiting relative frequency for the whole sequence. In terms of binary strings we could consider an infinite series of ones and zeros:

> "We say that it [the series] possesses the property of randomness if the relative frequency of the 1's (and therefore the 0's) tends to a certain limiting value which remains unchanged by the omission of a certain number of the elements and the construction of a new sequence from those which are left. The selection must be a so-called place selection, i.e., it must be made by means of a formula which states which elements in the original sequence are to be selected and retained and which discarded" [58].

A principle requirement is that the choice of the selection formula be made independently of the result of the corresponding observation, before anything is known about the result. For example, consider a binary string formed by a Bernoulli (binary) experiment (e.g. coin tossing):

10100101110011011010001011110001011010011011001101

The frequency of 1's in the string is $27/50 = 0.54$. By increasing the number sampled (conducting more trials) we may note that the relative frequency tends to 0.5. By using a formula that samples every odd element from the 50 samples above, we find the frequency of 1's is $14/25 = 0.56$. If, on the other hand we chose to sample elements at only prime number positions (2 3 5 7 11 13 17 19 23 29 31 37 41 43 47) the frequency of 1's is $9/15 = 0.60$. With strings larger than the 50 elements shown above, measured from the same experimental system, the limiting frequency of the 'odd' sampling method and the 'prime' sampling method would both tend to 0.5. Note that both the 'odd' and 'prime' methods can be chosen before knowing the exact result of conducting the measurement fifty times and obtaining the above string.

It is possible to choose a sampling method that would give a radically different relative frequency. Consider the case where we sample 15 elements from the above binary string at position numbers:

1,3,6,8,9,10,13,14,16,17,19,23,25,26,27.

In this case the relative frequency of 1's is $15/15 = 1.00$. It is also possible that selections be made to give any desired relative frequency. This is not a problem for von Mises' principle of randomness which only requires that the relative frequencies of selected subsequences converge on the whole sequence's limiting frequency as the subsequence lengths become at least denumerably infinite under the selection formula. However, the formulation of the selection criteria as described above would be ruled invalid, for von Mises maintains, under his definition of place selection, that the formula used for the selection of subsequences from infinite sequences "must leave an infinite number of retained elements and must not use the attributes of the selected elements, i.e., the fate of an element must not be affected by the value of its attribute." (op. cit. p.88) Assuming that the 50-element string above is just the first 50 elements of an infinite sequence, the selection of the 15 elements (1,3,6,8,9,10,13,14,16,17,19,23,25,26,27) to obtain a relative frequency of 1.0 violates both von Mises' conditions.

The existence of a limit to which relative frequencies converge is a big assumption, and though it appears to be borne out by vast quantities of empirical evidence from gaming and other sources, its existence is not guaranteed. Von Mises commissions the concept of a *collective*, a set of elements that gives rise to a 'mass phenomenon'. A collective is "a sequence of uniform events or processes which differ by certain observable attributes, say colours, numbers, or anything else" [58].

In early work von Mises acknowledges that, under his concept of randomness, proof of existence of a collective, in the analytic sense, is impossible.

> "A collective is completely determined by the distribution, i.e. by the (limits of the) relative frequencies for each attribute; it is however impossible to specify which elements have which attributes... [T]he existence of a collective cannot be proved by means of the actual analytical construction of a collective in a way similar, for example, to the proof of existence of continuous but nowhere differentiable functions, a proof which consists in actually writing down such a function" [59].

A collective that is truly random cannot be described by a formula or procedure. Von Mises clearly states this in the context of binary strings: "A sequence of zeros and ones which satisfies the principle of randomness cannot be described by a formula or by a rule such as: 'Each element whose place number is divisible by 3 has the attribute 1; all others the attribute 0'; or 'All elements with place numbers to squares of prime numbers plus 2 have the attribute 1, all others the attribute 0'; and so on" (ibid). Note here that the constraint of the impossibility of defining a function to represent a collective is directly related to the previous requirement of all infinite subsequences having the same limiting frequency as the entire sequence. If a function were available to describe the collective, then one could modify it to construct a subsequence that was comprised solely of 1's, or indeed any limiting relative frequency one desired.

An historical problem here arises for von Mises. At the time of his writing the influence of Logical Positivism was still being felt. Russell's theory of the predominance of description for existence and work by the British 'analysts' placed pressure on von Mises to justify existential status for an entity incapable of representation by explicit description or formula. Harold Jeffreys makes this criticism of von Mises definition of the collective:

> "The proof even of existence [of the collective] is impossible. On the limit definition, without some rule restricting the possible orders of occurrence, there might be no limit at all" [44].

In response to the analysts' criticism, von Mises' considers the possibility of restricting the definition of random sequences to a constructible subclass. Bernoulli sequences[28] have been shown [26] to be able to be constructed, hence, according to Jeffreys, existent. However this new definition of random sequences would prove too restrictive. Subsequence selection methods such as prime number selection would be precluded. In fact, limiting subsequence selection methods to any rule-based process would exclude some potentially legitimate subsequence due to the infinite number of selection methods. Instead, von Mises abandons attempts to placate critics by proving a collective's existence via formal description, and assumes an axiomatic approach, declaring that if we assume the aforementioned probability criteria hold for collectives, it can be shown[29] that selection of subsequences does not give rise to contradictions. He says, "Given a sequence of attributes, the assumption that the limits of the relative frequencies of the various attributes are insensitive to any finite or enumerably infinite set of place selections cannot lead to a contradiction in a theory based on this assumption" [58].

Further problems exist, however, for von Mises' definition of randomness. Alonzo Church maintains that it is too simplistic to be a rig-

[28] A Bernoulli sequence is one that is infinitely decomposable in a linear manner (i.e., pick every n^{th} element of the original sequence to form a sub-sequence, then offset by one and select new subsequence and repeat $n-1$ times) into sets of sub-sequences such that set of n sub-sequences has the following properties:
  Each of the sub-sequences has the same limiting frequency as the original;
  The sub-sequences are mutually independent,
  This is true for all n.

[29] This has been done by A.H. Copeland, A. Wald and W. Feller.

orous definition. Moreover it is possible to create a selection criterion that allows subsequences that violate the equality of relative frequency stipulations. Consider the following selection strategy ϕ proposed by Li and Vitányi, which generates subsequences from a collective:

> "Let $a = a_1 a_2 \ldots$ be any collective [of binary units]. Define ϕ_x as $\phi_x(a_1 \ldots a_{i-1} = 1$ if $a_i = 1$ and undefined otherwise. But then [limiting relative frequency] $p = 1$. Defining ϕ_w by $\phi_w(a_1 \ldots a_{i-1}) = b_i$, with b_i the complement of a_i, for all i, we obtain ... $p = 0$. Consequently if we allow functions like ϕ_x and ϕ_w as strategy, then von Mises definition cannot be satisfied at all" [55].

Von Mises, in later work, was aware of this shortcoming through criticisms by Abraham Wald and others and accepted restriction of the class of permissible place selection strategies:

> "Given a sequence of attributes, the assumption that the limits of relative frequencies of the various attributes are insensitive to any finite or enumerably infinite set of place selections cannot lead to a contradiction in a theory based on this assumption. It is not necessary to specify the type or properties of the place selection under consideration. It can be shown that, whatever enumerably infinite set of place selections is used, there exists sequences of attributes which satisfy the postulate of insensitivity" [58].

It was Wald [83] who originally proposed a solution to the problem of place selection by restricting the set of a priori admissible strategies to a fixed countable set of functions. But which set? Church suggests that the set should be precisely those strategies that are computable in accordance with his thesis.[30] This means that the set of ϕ is the set of recursive functions.[31] This approach has the additional advantage of the rigorous definition associated with recursive functions. This amended version of von Mises' definition of randomness is call Mises-Wald-Church randomness and can be summarized as follows. A binary string is Mises-Wald-Church random if and only if:

---

[30] Church's Thesis states that a function is computable if it is computable by a Turing machine, or more precisely: The intuitively and formally defined class of effectively computable partial functions coincide exactly with the class of functions that are computable by an unlimited register machine (URM).

[31] It is provable that a function is *recursive* if a Turing Machine running that function eventually halts for all inputs for that function.

1. The relative frequency of all attributes of a collective approaches a definite value between 0 and 1 in the limit as the collection length approaches infinity;
2. Every subsequence chosen according to some admissible place selection function has the same limiting relative frequency as the entire collection;
3. A place selection function is admissible if and only if it is a member of the set of recursive functions.

However, Mises-Wald-Church randomness still fails to fully capture the concept of randomness as examples of sequences were found that were intuitively non-random but met the Mises-Wald-Church criteria. In particular, it was shown by Ville [82], that there exist binary sequences which satisfy the Mises-Wald-Church definition, have a limiting relative frequency of the occurrence of ones equal to 0.5 and have the property

$$\frac{f_n}{n} \geq \frac{1}{2} \text{ for all } n.$$

As Li and Vitányi note,

> "The probability of such a sequence of outcomes in random flips of a fair coin is zero. Intuition: if you bet '1' all the time against such a sequence of outcomes, then your accumulated gain is always positive. Similarly, other properties of randomness in probability theory such as the Law of the Iterated Logarithm[32] do not follow from the Mises-Wald-Church definition" [55].

One method of avoiding the inclusion of these problematic strings is to maintain that all random strings not only pass one test of randomness, such as normality, but *all* possible tests. This method, based on constructive measure theory and proposed by P. Martin-Löf [57], is formulated as follows. We define a string to be 'random' if it is 'typical'. A string is typical just in case it is not 'special', that is, it does not possess any particular distinguishing property such as, for example, being an infinite binary string with a finite number of ones. Typical infinite strings have the converse property: they will have an infinite number of ones. In fact we require that typical strings will have all converse

[32] The Central Limit Theorem provides a converging estimator for an attribute of a collective as n proceeds to infinity, but it says nothing about the fluctuations of the estimator around its true value. The Law of the Iterated Logarithm complements the Central Limit Theorem by describing extreme bounds for these fluctuations and shows that they are of the same order of magnitude as $2loglog(n)^{1/2}$.

properties. Thus a typical string will belong to every reasonable statistical majority. So now we can say that an element (string) x will be "random" just in case x lies in the intersection of all such majorities. Li and Vitányi note,

> "Suppose that a single particular property, such as containing infinitely many occurrences of ones (or zeros), the Law of Large Numbers, or the Law of the Iterated Logarithm, has been shown to have probability one, then one calls this a *Law of Randomness.* Each sequence satisfying this property belongs to a large majority, namely the set of all sequences satisfying the property which has measure one by our assumption.
>
> "Now we call an infinite sequence 'typical' or 'random' if it belongs to *all majorities of measure one*, that is, it satisfies all Laws of Randomness. In other words, each *single* individual 'random' infinite sequence possesses all properties which hold with probability one for the ensemble of all infinite sequences" (ibid).

However, at this point a difficulty arises. All complements of singleton sets have probability one, thus the intersection of all sets of probability one is empty. This implies there are no random strings. Martin-Löf overcame this problem by employing Turing's premise that every effective[33] mathematical method can be carried out by the universal Turing machine. Martin-Löf argued that a law of probability be a partial recursive function.[34] By doing this he could define the set of random infinite sequences, not as the intersection of all sets of measure one, but rather as the intersection of all sets of measure one *with a recursively enumerable complement.*[35] This intersection is itself a set of measure one with a recursively enumerable complement. The result of this is that we have one single effective law of randomness: random strings satisfy all effective laws of randomness.

[33] 'Effective' here is used in technical sense as proposed by Church. A method is 'effective' if it can be realised by a human clerk with paper and a pencil in arbitrary but finite time.

[34] A partial recursive function is a function computed by a Turing machine that need not halt for all inputs. (A partial function is a function which is not defined for some of its domain).

[35] A set of integers is said to be recursively enumerable if it constitutes the range of a general recursive function. A general recursive function is a function that when computed is defined for all arguments. That is, when it is computed on a Turing machine, the machine will halt for all inputs.

In terms of my development of a unifying theory of information, an important aspect in Martin-Löf's formulation is the notion of 'typicality'. He requires that random strings not possess any property capable of distinguishing them from other strings. I will develop a model of distinguishability in section 3.1, based on Leibniz's theory of indiscernibles, and show how it forms the conceptual foundation of information.

The development of Martin-Löf's definition of randomness was strongly influenced by Andrei Kolmogorov. Martin-Löf met Kolmogorov on visit to Moscow during 1964–1965 while studying the complexity oscillations of infinite sequences. Kolmogorov had, in 1933, published "Foundations of Probability" and at the time of Martin-Löf's visit, had moved to working on an Algorithmic Information Theory designed to overcome problems with probabilistic accounts of information. The next section details Kolmogorov's algorithmic theory of information.

Kolmogorov

Kolmogorov directly considered the issue of the quantification of information. In his original paper [48], Kolmogorov examined the two existing approaches to the quantification of information: the combinatorial approach (Brillouin, Szilard) and the probabilistic approach (Shannon), then proposed a third approach – the algorithmic approach – to overcome perceived short comings in the previous two.

Kolmogorov states: "Actually it is most fruitful to discuss the quantity of information 'conveyed by an object' (x) "about an object" (y)" (Kolmogorov 1965, p.4). Probabilistic information theory for continuous distributions presents the generalized form:

$$I(x,y) = \iint P_{xy}(dxdy)\log_2(\frac{P_{xy}(dxdy)}{P_x(dx)P_y(dy)})$$

The form of this equation reflects the Shannon type of information

$$H = \int_{-\infty}^{\infty} p(x)\log p(x)dx,$$

augmented to the case of conditional mutual information using a chain rule. In practice, we are interested in relatively simple x, y relationships. That is, the informationally relevant mappings between x and y are a small subset of the entire set of attributes of the objects. For example, Kolmogorov says: "While a map yields a considerable amount of information about a region of the earth's surface, the microstructure of the paper and the ink on the paper have no relation to the microstructure of the area shown on the map" (ibid).

It is meaningless, Kolmogorov tells us, to ask what information the string '0 1 1 0' could convey about the sequence '1 1 0 0'. However, if we take a random number (from a specified statistical table) and for each of its digits calculate a sequence of single digit integers by taking the last digit (in the units place) of the square of the random number, the sequence will contain approximately $(\log_2 10 - 0.8)n$ bits of information about the original sequence (n = the number of digits in the original sequence). That is, each of the digits in the top row of the following table generates a new sequence using the corresponding digit in the bottom row.

0 1 2 3 4 5 6 7 8 9

0 1 4 9 6 5 6 9 4 1

Thus, 7 8 4 7 5 0 1 generates 9 4 6 9 5 0 1. The reason that the generated sequence holds $(\log_2 10 - 0.8)n$ bits of information about the original string is because any patterns or regularities in the original string will be at least partially preserved in the mapping process due to the mapping regularities. That is, 1 in the generated string will always map back to 1 or 9, 9 to a 3 or 7.[36]

Generalizing this, Kolmogorov attempts to define,

$$I_A(x : y),$$

that is the information that x tells us about y. Kolmogorov considers a set X of enumerable objects x. Let $n(x)$ be a sequence of natural numbers represented by ones and zeros that acts as an index for X. We denote the length of the sequence $n(x)$ by $l(x)$. Now we make the following assumptions[37]:

1. D is the set of all $n(x)$ and the correspondence of X to D is one to one
2. $D \subset X$, the function $n(x)$ on D is generally recursive, and for $x \in$ D,

$$l(n(x)) \leq (x) + C,$$

where $C$is a constant.

[36] It is interesting to note the reflective symmetry in the bottom row!

[37] Kolmogorov notes that not all these assumptions are essential but they simplify the discussion.

3. $x \in X$ and $y \in X$. X also contains the ordered pair (x, y) whose index is a generally recursive function of the indices of x and y and that,

$$l(x,y) \leq (y) + C_x,$$

where C_x is dependent solely on x.

Now it is possible that $n(y)$ may not be the most efficient way of specifying y given x so we consider the generation of string y from x by means of a program p which can be defined by the function:

$$\varphi(p, x) = y$$

that associates object x and program p with object y. We assume φ is partially recursive. Now we can take as the "relative complexity" of object y given x, $K\varphi(y|x)$, the minimal length $l(p)$, achieved when p is of minimal length.

$$K_\varphi(y|x) = \begin{cases} \min_{\varphi(p.x)=y} l(p) \\ \infty, \text{ If there is no p such that } \varphi(\text{p.x}) = \text{y} \end{cases}$$

A function $v = \varphi(u)$ of $u \in X$ with range $v \in X$ is said to be partially recursive if it generates a partially recursive function of the index transformation $n(v) = \psi[n(u))]$. Kolmogorov says,

> "In order to understand the definition, it is important to note that, in general, partially recursive functions are not defined everywhere, and there is no fixed method for determining whether application of the program p to an object k will lead to a result or not. As a result, the function cannot be effectively calculated (generally recursive) even if it is known to be finite for all x and y" [48].

Kolmogorov proves what he calls the "Fundamental Theorem" which states that there exists a partially recursive function $A(p, x)$ such that for any other partially recursive function $\varphi(p, x)$:

$$K_A(y|x) \leq gK_\varphi(y|x) + C_\varphi$$

where C_φ is independent of x and y. Functions that satisfy the Fundamental Theorem are called "asymptotically optimal" and their corresponding complexity $K_A(y|x)$ is finite. In the final step to defining $I_A(x{:}y)$ Kolmogorov notes that

$$K_A(y) = K_A(y|1),$$

and that we can define the quantity of information conveyed by x about y as

$$I_A(x:y) = K_A(y) - K_A(y|x).$$

Kolmogorov originally (as presented here) was concerned with the mutual information in two objects. However the model he developed is commonly used to specify the intrinsic information of an individual object. This is notionally equivalent to a program p generating a string y without separate input. Here we have:

$$K_\varphi(y) = \begin{cases} \min_{\varphi(p)=y} l(p) \\ \infty, \text{ If there is no p such that } \varphi(\text{p}) = \text{y} \end{cases}$$

Kolmogorov notes that his method has one important disadvantage: there is no provision made for determining the shortest possible p. Kolmogorov states that the approach he developed does not,

> "allow for the 'difficulty' of preparing a program p for passing from an object x to an object y. By introducing appropriate definitions, it is possible to prove rigorously formulated mathematical propositions that can be legitimately interpreted as an indication of the existence of cases in which an object permitting a very simple program, i.e., with a very small complexity K(x), can be restored by short programs only as the result of computations of a thoroughly unreal duration" [48].

Since Kolmogorov's reasoning is somewhat abstract and primarily focuses on the conditional case of information, it is worthwhile to summarise his analysis and look at the absolute case. The easiest way of thinking about Kolmogorov information is to imagine giving instructions to another person to generate a specific string. If you can do this unambiguously so that the other person generates precisely the required string in finite time then the number of bits in that instruction represents an upper limit on the information content of the string. The string will contain at most that much Kolmogorov information. For example, if the instruction is "Print the first 1,000,000,000 bits of π." then, assuming 8 bits per symbol (ASCII), an upper limit of the information of that string is 8x40= 320 bits. There may exist shorter instructions, but any longer instructions will contain superfluity.

This means the information content of a string, y, is the minimum length of an instruction or program, p, to be executed by a universal computer, φ (in the above example, a person) if there exists a program that generates that string, and infinite otherwise.

In turns out, as an extension of Martin-Löf's work, that most 1,000,000,000 bit sequences have a Kolmogorov Information value of about 1,000,000,000 bits. This is because they are random and the shortest instruction to generate the string is to print the string itself. This is of the form "Print 01010111010100101001..." which is approximately a billion bits long. The conditional mutual information form, that was developed in Kolmogorov's paper, may be thought of as the unambiguous specification of the generation of a sequence given another sequence as input.

A few years before Kolmogorov published his paper on the quantitative definition of information, R.J. Solomonoff was working towards a similar outcome starting from a different point. Solomonoff contributed to the core foundations for Algorithmic Information Theory in 1960 in his paper *A Preliminary Report on a General Theory of Inductive Inference* in which tried to overcome problems associated with Carnap's theory of inductive reasoning [16]. The next section examines his contribution.

Solomonoff

Solomonoff examines the problem of inductive inference given initial information by considering the problem of predicting future occurrences of string elements given an initial sequence of a substring. That is, given a long series of symbols represented by T, what is the probability that it will be followed by a subsequence represented by an element a? What is $p(a|\,T)$? Carnap's approach was to predict the outcome using priors[38] generated by taking a weighted sum of all explanations in a given description language. Solomonoff developed a similar approach, but whereas Carnap's method was appropriate for only the simplest finite languages, Solomonoff's theory was more generalized, applying to a universal description language for Turing machines. At the end of the development of this method, Solomonoff arrived at a final equation (his Equation (5)) that returns a combinatorial probability of a sequence occurring based on the number of ways that it could have been formed. Solomonoff states,

> "Consider all possible sequences of symbols that could be descriptions of all the things a person might observe in his life. These sequences correspond to the sequences being encoded in Equation (5) ...

[38] That is, prior judgments about parameters and about the chances of particular outcomes.

> A complete model that 'explains' all regularities observed in these sequences is that they were produced by some arbitrary universal machine with a random binary sequence as its input. Equation (5) then enables us to use this model to obtain prior probabilities to be used in computation of posterior probabilities using Bayes' Theorem. Equation (5) finds the probability of a particular sequence by summing the probabilities of all possible ways in which that sequence might have been created" [76].

Importantly for Algorithmic Information Theory, Solomonoff introduces the concept of a binary description:

> "Suppose that we have a general purpose digital computer M_1 with a very large memory. ... Any finite string of 0's and 1's is an acceptable input to M_1. The output of M_1 (when it has an output) will be a (usually different) string of symbols, usually in an alphabet other than the binary. If the input string S to machine M_1 gives output string T, we shall write
>
> $$M_1(S) = T$$
>
> Under these conditions, we will say that 'S is a description of T with respect to machine M_1.' If S is the shortest such description of T, and S contains N digits, then we will assign to the string, T, the *a priori* probability 2^{-N}" (ibid).

Solomonoff's intention was to create a formalised theory of scientific inductive reasoning. Under Solomonoff's model, observations could be represented as binary strings. Scientific theories intended to explain the observations can be described as algorithms which produce those strings. These theories will be of varying complexity and hence their corresponding algorithms will vary in size. The scientist would apply Occam's Razor to select between the completing theories and choose that theory whose algorithm has the shortest length. If the observations made were of a random event (or if there were a great degree of inherent randomness in the observation process) no theory would be effective in describing the observation.

The similarity to Kolmogorov's work is evident. Solomonoff's contribution to the field led eventually the development of feasible inference methodologies such as the Minimum Description Length[39] (MDL) principle in statistical modelling.

[39] MDL Principle states that any regularity in a given set of data can be used to *compress* the data, i.e. to describe it using fewer symbols than needed to describe the data literally.

About the same time as Kolmogorov and Solomonoff were developing their accounts of algorithmic information, Gregory Chaitin independently developed the notion of Kolmogorov information measure in the mid-nineteen-sixties. As Chaitin states, "This definition was independently proposed about 1965 by A. N. Kolmogorov of the Academy of Science of the U.S.S.R. and by me, when I was an undergraduate at the City College of the City University of New York. Both Kolmogorov and I were then unaware of related proposals made in 1960 by Ray J. Solomonoff of the Zator Company in an endeavor to measure the simplicity of scientific theories" [19].

We will look at Chaitin's work in the next section.

Chaitin

The most important of Chaitin's early contributions to the formulation of algorithm complexity theory emerged from his study of randomness. Capturing the same essence as Solomonoff and Kolmogorov and furthering the work of Martin-Löf, Chaitin determined that a "random" string is one that cannot be algorithmically compressed .

In Chaitin's earliest work he considers, in a manner very similar to Solomonoff's analysis of inductive inference, the search for regularity in scientific data:

> "Consider a scientist who has been observing a closed system that once every second either emits a ray of light or does not. He summarizes his observations in a sequence of 0's and 1's in which a zero represents 'ray not emitted' and a one represents "ray emitted." The sequence may start
>
> 0110101110 ...
>
> and continue for a few thousand more bits. The scientist then examines the sequence in the hope of observing some kind of pattern or law. What does he mean by this? It seems plausible that a sequence of 0's and 1's is patternless if there is no better way to calculate it than just by writing it all out at once from a table giving the whole sequence:
>
> My Scientific Theory: 0
> 1
> 1
> 0
> 1

0
1
1
1
0

> This would not be considered acceptable. On the other hand, if the scientist should hit upon a method by which the whole sequence could be calculated by a computer whose program is short compared with the sequence, he would certainly not consider the sequence to be entirely patternless or random. And the shorter the program, the greater the pattern he might ascribe to the sequence" [18].

To develop a formal description of this concept, Chaitin considers a Turing Machine that produces a binary string output (Chaitin, 1969). Let S be the set of all binary strings. Define L as follows: An N-state, 3-tape-symbol Turing machine[40] can be programmed to produce S iff $N \geq L(S)$. Chaitin notes the use of the letter "L" is suggested by the phrase "the *L*ength of program necessary for computing S". This definition captures the fact that the length of a program run on this machine to output $S$is less than or equal to the total number of distinct states of the machine. Given L, Chaitin defines a subset of S, C_n, as follows:

$$L(C_n) = \max_{\text{S of length n}} L(S)$$

The maximum here is taken of all binary strings S of length n. The symbol C_n then conceptually denotes the most complex binary sequences of a length n. Based on his philosophical considerations noted above (that a random string is one that cannot be represented by a program shorter than the string itself), Chaitin argues that for random S of length n:

$$L(S) \approx L(C_n)$$

Chaitin does not provide a rigorous deductive proof of this, however he offers formal inductive credence by "proving various results concerning what may be termed statistical properties of such finite binary sequences" [20]. These properties include familiar tests for randomness such as Simple Normality and Von Mises Place Selection. For each of these tests he shows the above relationship holds.

[40] That is, a Turing machine that uses tape that has 3 symbols: 0, 1, <space>.

In the case of infinite sequences, Chaitin defines the set of infinite random binary sequences, C_n, as a subset of the set of all infinite binary sequences, S, that satisfy for all sufficiently large values of n the following:

$$L(_n) > L(C_n) - f(n)$$

where $f(n) = 3log_2n$ (for a 3-tape-symbol Turing machine) and S_n = the first n bits of S (Chaitin, 1966). Chaitin admits the definition is somewhat arbitrary, "The failure to state the exact cut-off point at which $L(S)$ becomes too small for S to be considered random or patternless gives us the first definition of its informal character. But in the case of finite binary sequences, no gain in clarity is achieved by arbitrarily settling on a cut-off point, while the opposite is true for infinite sequences" [20].

Chaitin's work on random strings is essentially complementary to the work done by Kolmogorov and Solomonoff. To contrast Chaitin's approach with Kolmogorov's $K\varphi$ theorem we turn to Chaitin's general formulation with respect to binary computing machines. Chaitin proposes the following model,

"Formally, a binary computing machine is a partial recursive function M of the finite binary sequences which is finite binary sequence valued. The argument of M is the program $[P]$, and the partial recursive function gives the output (if any) resulting from that program. $L_M(S)$ and $L_M(C_n)$ (if the computing machine is understood, the subscript $[M]$ will be omitted) are defined as follows:

$$L_m(S) = \begin{cases} \min_{M(P)=S}(\text{Length of P}) \\ \infty, \text{ If there is no such P} \end{cases}$$

$$L_m(C_n) = \max_{\text{S of length n}} L_m(S)$$

In this general setting the program for the definition of the random or patternless binary sequence assumes the following form: The patternless or random finite binary sequences of length n are those sequences S for which $L(S)$ is approximately equal to $L(C_n)$. The patternless or random infinite binary sequences S are those whose truncations S_n are all patternless or random finite sequences. That is, it is neccesary that for large values of n, $L(S) > L(C_n) - f(n)$ where f approches infinity slowly" [20]. We see here that $L(S)$ is in essense Kolmogorov's $K_\varphi(y)$

In summary, Chaitin's basic premise is that random sequences do not have regularities; they are, as he frequently tells us, "patternless". This means there are no redundancies in the sequence that may be exploited in the generation of the string, thus the length of a program which generates a finite random string must be approximately the length of the string itself. A string is random just in case it cannot be algorithmically compressed. In informatic terms, this means the total information in the string is the string itself. In this way in the set of sequences of length n, the set of random sequences C_n will be those strings which possess the most information for strings of that length. This is an important realisation because it appears somewhat counter-intuitive that random entities are informatically maximal. We will be examining this significant relationship in detail in section 4.7.

2.2.4 Signpost

We have reached the end of our survey of the three main approaches to a formalisation of information: the three entrances, starting from different points, into the mine of information. The survey has been technical and at times abstract. So it is worthwhile at this stage to summarise the approaches and highlight the essential issues, and since the three have some aspects in common, it is also worth examining the relationship between them as well as their relationship with information.

The Thermodynamic/Statistical Mechanical approach to information essentially establishes the relationship between entropy and information. That relationship, Brillouin and Schrödinger tell us, is that of reversed sign: information is negative entropy. Thus in our search for a foundational account of information, the question becomes: what is the nature of entropy? Boltzmann-Maxwell Statistical Mechanics informs us that entropy is a combinatorial property determined by the number of possible distinct macrostates which a system may exhibit. We assume that all of these states are equally likely to manifest so that the probability of any particular one appearing is the inverse of the total number of states. Taking the logarithm of this probability (and multiplying by an appropriate constant) returns a quantity of the same absolute value as the system entropy but of opposite sign. What does this mean? Entropy represents the capacity of a system to manifest a certain number of distinct states. The more physical states the system can manifest, the greater the entropy. Information is the distinguishing of one of those particular states.[41] It is the *selection* of one unique member from the total set. In essence, entropy is the count of unresolved physical possibilities of a particular system, while information

[41] We could distinguish 'one or more' states. In that case we would be calculating the information of that particular sub-set. For simplicity's sake we will distinguish just one.

is the resolution of this system to a particular state; it is the collapse of potentiality to actual selection. I have previously noted that 'distinguishability' is crucial to defining information and we will see later that Boltzmann's account of entropy requires a bit more work to be compatible with entropy observations specifically due to problems of distinguishability.

Other important issues concerning information have arisen from our considerations of the Thermodynamic/Statistical Mechanical approach. We have seen that information is a very real quantity with the ability to produce real work. The sorting process can be used to generate thermodynamic work. We will examine this notion in more detail in Section 4.6.3 when we examine Gibbs' Paradox and its relationship to information. I wish to emphasise the capacity of informatic processes (such as sorting) to produce work in order to dispel any notion that information is anything other than a real, objective quantity.

We have also seen that the related concepts of measurement and memory are important in resolving the Maxwell's Demon paradox which relates entropy and information. Szilard's analysis of Maxwell's Demon notes the importance of the role of measurement. The capacity to measure the properties of a particle and distinguish it from particles with other property values prevents the Demon violating the Second Law. Szilard notes that the "measurement procedure is fundamentally associated with a certain definite average entropy production, and that this restores concordance with the Second Law" [79]. In Section 3.2, the apprehension of information from an informatic object is detailed and Section 4.5.2 further discusses measurement, Demons and information.

Memory is significant by virtue of its role in resetting the system so that a work cycle could operate. Memory is also central to the concept of distinguishability of consecutive temporal states of a system, which we will develop in Section 3.1. The relationship between information and memory is discussed in Section 4.2.2.

Shannon's theory of information is closely related to Boltzmann-Maxwell entropy, though the two accounts differ in one significant way. Shannon is not directly concerned with the nature of the ensemble of possible messages that can manifest because he makes no assumptions about equiprobability of messages. What he develops is a more general case where he considers the individual probability for each possible message that may be sent. The probability distribution need not be uniform. The case where all messages are equally likely is just a special case.

The algorithmic approach to information is considerably different from the other two approaches. Here there is no ensemble of unresolved possibilities, so there cannot rightly be said to exist entropy. The algorithmic technique is generative in nature and calculates the minimum amount of code required to create a specific string (or 'message' in Shannon's parlance). This minimum code quantifies the information associated with that string. How then, if at all, is this related to Boltzmann-Maxwell and Shannon types of information? If all three theories refer to the same thing when they talk about information, what is the primal quantity that, when viewed from different conceptual angles, can be described by these different accounts? The unification of these theories of information is a key goal of this book and an attempt at this goal is developed in sections 3 and 4. It will suffice at this point to say that they are all subsumed by the concept of symmetry.

Before proceeding to develop a unifying account of information we should tie up a conceptual loose end. All three theories of information that we have examined so far have made reference to a quantity called 'probability'. It is important to further investigate this quantity in order to correctly determine its place in a unified theory of information.

2.3 Probability

We have seen in the previous sections that Statistical Mechanics and all other accounts of information rely on probabilistic accounts of system state. Maxwell and Boltzmann constructed momentum probability distributions to give a microscopic underpinning to the macroscopic quantity entropy. Brillouin talks of probabilities of possible outcomes changing when information is received. Shannon considers information being generated by a Markov Process creating a sequence of symbols governed by a set of probabilities. Von Mises judges probability to be a primary physical property of a set, "a physical constant belonging to the experiment as a whole and comparable with all its other physical properties" [58]. Solomonoff speaks of finding "the probability of a particular sequence by summing the probabilities of all possible ways in which that sequence might have been created" (Solomonoff, 1960, p.11). We have already noted, Kolmogorov states "the applicability of the results of the mathematical theory of probability to real 'random phenomena' must depend on some form of the frequency concept of probability" (Kolmogorov, 1963, p.387). Chaitin speaks of "the probability that [a Turing machine] M eventually halts with the string s written on its output tape"[21]. But, what *is* probability? Is Shannon's notion of probability the same as Boltzmann's, or Chaitin's? If not, how can we compare them? In this section I will examine several possible meanings of the term *probability* in the context of information theory and offer a working definition that can be used to understand all three approaches studied thus far.

2.3.1 Subjective Probability

The subjectivist account of probability maintains that probability is solely a knowledge phenomenon, that it is an *epistemological* rather than an *ontological* issue. On this account randomness is not objective and measurable but represents our lack of knowledge. If we were perfectly informed of every single parameter concerning a coin toss, we would be able to nominate with absolute certainty the outcome of the event.

The notion that probability is semantically grounded in beliefs is an old one. Laplace maintains that probability is a measure in part of our ignorance concerning an event and in part of our knowledge of that event [50]. Bayes also held this doxastic interpretation and examined the relationship between probabilities and personal hypotheses. The probability $P(H|E)$ of a hypothesis given evidence E is the degree of belief in the veracity of H corroborated by E. Bayes' Theorem states that given a priori probability of H, $P(H)$, and the conditional probability of E given H, the probability of H given E is by:

$$P(H|E) \propto (H)P(E|H).$$

It is clear that the subjective account of probability is primarily concerned with changing probability assignments on presentation of new evidence.

2.3.2 Frequency Probability

The frequentist interpretation of the notion of probability is essentially that probability only has meaning as a result determined by experimental outcomes. 'Classical' frequentist definitions of probability, such as J. Neyman maintain that probability is defined as follows: If there are n possible alternatives of which there are m where proposition p is true then the probability of p is m/n [62]. Most influential of these was von Mises. Von Mises' goal was, like Kant's, to synthetically, rather than analytically, define the word *probability.* This was to be done in an environment where many definitions, 'intuitive' and technical, currently exist. To invoke an analogy, von Mises quotes Werner Sombart's book *Practical Socialism* in which Sombart tries to define socialism amidst a number of existing interpretations. Sombart concludes,

> "The only remaining possibility is to consider socialism as an idea and to form a reasonable concept of it, i.e. to delimit a subject matter which possesses a number of characteristics considered to be particularly important to it and which form a meaningful unity; the 'correctness' of this concept can only be judged from its fruitfulness, both as a creative idea in life and as a useful instrument for advancing scientific investigation." [58].

Von Mises' desired definition of probability must firstly be independent of popular current usage and secondly be a concept that is not gauged by its "correspondence with some usual group of notions, but only by its usefulness for further scientific development, and so, indirectly for everyday affairs"[58]. Von Mises leaves open the possibility of finding an analytic definition of probability.

To illustrate how a definition of probability may become sidetracked, von Mises considers other scientific notions, such as *work*, which have everyday currency. Scientifically, work is the scalar product of the vectors of force and displacement. In popular idiom, it is common to refer to work done in book-writing, practicing a musical instrument or performing surgery. While referencing exertion, the meaning of these phrases is far from the scientific definition above. Although popular usage of the term *work* may not pose difficulties for the scientific definition, the problem is potentially more serious when considering *probability*.

It is common to talk of the probability of rain this evening or the probability of an outbreak of war between two countries. This, von Mises argues, is meaningless with respect to his theory of probability. Employments such as these must be eliminated from the scope of the definition. Probability, he maintains, is a mass property: it does not apply to individual events. To develop this case, von Mises offers three examples,

1. What is the probability that a double 6 will appear x times if two dice are thrown y times?
2. What is the probability that x men out of a population y will die aged z years?
3. What is the probability that the mean molecular velocity will be greater than x in a specified closed system?

Central to von Mises' theory of probability is the concept of *the collective*. In the three examples above, each has mass character, applying either to a large population or an unlimited repetition, which distinguishes it from the single-event colloquial use of probability noted earlier. The term 'the collective', states von Mises, denotes,

> "a sequence of uniform events or processes which differ by certain observable attributes, say colours, numbers, or anything else... All the throws of the dice made in the course of a game form a collective wherein the attribute of the single event is the number of points thrown. Again, all the molecules in a given volume of gas may be considered as a collective, and the attribute of a single molecule might be its velocity. A further example of a collective is the whole class of men and women whose ages at death have been registered by an insurance office" [58].

Given these definitions of the collective and of attributes, von Mises' definition of probability then is concerned solely with 'the probability of encountering a certain attribute in a given collective' (ibid.). The quantification of this probability remains to be defined. Von Mises examines the results of measuring attributes in repeated experiments on a collective. For example, if two dice are rolled 200 times we may find that double 6 occurs 5 times. The relative frequency of this occurrence is $5/200 = 1/40$. If the collective had a size of 1800, that is if the dice experiment were performed 1800 times, then we might find that the relative frequency is $48/1800 = 1/37.5$, or perhaps $60/1800 = 1/30$. If we continue to increase the number of experiments we would notice that, if we were to plot the relative frequency against the number of experiments, after some large initial oscillations the relative frequency would tend toward a specific value as the collective size increases. Eventually we would reach a collective size above which the oscillations around that specific value are indistinguishable.[42] This value is known as the limiting value of the oscillations. In the dice example above, the limiting value is $1/36$.

With the addition of limiting frequency value, von Mises hones his definition of the collective:

> "We will say that a collective is a mass phenomenon or a repeated event, or, simply, a long sequence of observations for which there are sufficient reasons to believe that the relative

[42] This is Poisson's familiar Law of Large Numbers, though von Mises is initially reluctant to call it such due to ambiguity in Poisson's use of the law of this name.

> frequency of the observed attribute would tend to a fixed limit if the observations were indefinitely continued" [58].

The next step is to equate the probability of an attribute with the limit that its relative frequency of occurrence approaches. "This limit will be called *the probability of the attribute considered within the given collective*" (ibid.).

It is crucial to understand that the limiting frequency is a physical property of the system and not just an experimental artefact:

> "Here we have the 'primary phenomenon' (Urphänomen) of the theory of probability in its simplest form. The probability of a 6 is a physical property of a given die and is a property analogous to its mass, specific heat, or electrical resistance. Similarly, for a given pair of dice (including of course the total set-up) the probability of a 'double 6' is a characteristic property, a physical constant belonging to the experiment as a whole and comparable with all its other physical properties. The theory of probability is only concerned with relations existing between physical quantities of this kind" [58].

Here von Mises rejects any subjective interpretation of probability. Probability is an objective, physical property of a dynamic mass system, the value of which is to be obtained by experimental determination of limiting frequency through measurement of relative frequencies.

Von Mises maintains that the only legitimate scientific use of probability is with reference to a collective. As mentioned above, isolated events have no probabilistic significance. By way of illustration, consider death rates as used by the insurance industry. The collective may be defined as all men insured before their fortieth birthday with the attribute dying in their forty-first year. Say, now, that the limiting frequency for this collective is p. It is meaningless to say that the probability of a particular male in his forty-first year, Mr. X, has the probability p of dying because p only has semantic significance for the defined collective. Mr. X is a member of a great number of collectives (e.g. people who work in profession A, people who live in city Q) each with its own limiting frequency for the attribute of death in the forty-first year. No one of these is *the* probability of Mr. X dying. For instance, Mr. X is a member of the total population. Since the age-death rate for women is lower than that for men, the limiting frequency of the attribute 'death in the forty-first year' for the collective formed by the entire population will typically be $q < p$. What then is the probability of Mr. X dying in his forty-first year: q or p? One may argue that the rational approach would be to take the most restrictive collective possible (beings who are human, male, work in profession A, live in city Q, etc.) and use that frequency to define the attribute probability. However, that approach fails because the most restrictive collective possible will be a singleton with Mr. X as its only member, which is no collective at all.

Laplace considers the notion of *extraordinary* outcomes, results that seem so unlikely that some inquiry into the causal process might be justified. These results may occur in the drawing of letters in a scrabble game or numbers in a lottery.

> "On a table we see letters arranged in this order *C o n s t a n t i n o p l e*, and we judge that this arrangement is not the result of chance, not because it is less possible than the others, for if this word were not employed in any language we should not suspect it came from any particular cause, but this word being in use among us, it is incomparably more probable that some person has arranged the afore said letters than that this arrangement is due to chance" [50].

We know that, in a million-number lottery, drawing the number 400,000 is just as likely as, say, 475,813. Yet intuitively it seems utterly improbable that *C o n s t a n t i n o p l e* would appear from a random draw or that ticket 400,000 would win the lottery. Von Mises' resolution to this 'paradox' rests in the use of collectives. In Laplace's example, the event that the sequence *C o n s t a n t i n o p l e* would be picked has, in itself, no 'probability' at all as it is an isolated event. One can compose a collective of the 26^{14} possible permutations of fourteen-character-long strings of letters in the English language and assign attributes 'meaningless' and 'coherent' to the members of the collective. The number of 'meaningless' strings greatly outnumber the 'coherent' ones which is why *C o n s t a n t i n o p l e* seems unlikely. The string *m o s t i m p r o b a b l e* is equally likely and equally extraordinary. The same principle applies to the lottery numbers. We would find the numbers 200,000 or 300,000 just as unlikely as 400,000. A collective could be formed of repeated lottery draws and the attributes chosen as numbers that are round hundreds of thousands and those that are not. The limiting frequency for those numbers that end in five zeros will be almost 100,000 times smaller than for those that do not.

These considerations of extraordinary outcomes indirectly lead to notions of pattern and regularity. One of the reasons that we instinctively believe that ticket 475,813 is a more believable lottery winner than ticket 400,000 is that the number 400,000 is so 'regular'; it does not appear random. As we have seen, von Mises considered the notion of randomness, and his work was seminal in creating what would eventually become Algorithmic Information Theory. For von Mises, randomness, when applied to his collective model of probability, was strongly based on the idea of place selection and the possibility of selecting elements from the collective in such a manner as to change the limiting frequency. If it is possible, according to von Mises, to select by means of some rule, a finite or an at least denumerably infinite subsequence from the total collective and an attribute's limiting frequency in the subsequence is different to that of the collective, then the collective cannot be said to be random. The motivation is that the probability of the attribute is a physical property of the collective and should always remain invariant under data partitioning. This concept of randomness is important for the definition of a collective. Von Mises says:

> "A collective appropriate for the application of the theory of probability must fulfil two conditions. First the relative frequencies of the attributes must possess limiting values. Second, these limiting value must remain the same in all partial sequences which may be selected from the original one in an arbitrary way. Of course, only such partial sequences can be taken into consideration as can be extended indefinitely, in the same way as the original sequence itself" [58].

In this quote we see an early incarnation of E.T. Jaynes' Maximum Entropy Principle which we will visit later (Section 4.4.1).

Thus we arrive at a summary of von Mises' definition of probability:

> "1. It is possible to speak about probabilities only in reference to a properly defined collective.
>
> 2. A collective is a mass phenomenon or an unlimited sequence of observations fulfilling the following two conditions: (i) the relative frequencies of particular attributes with the collective tend to fixed limits; (ii) these fixed limits are not affected by any place selection. That is to say, if we calculate the relative frequency of some attribute not in the original sequence, but in a partial set, selected according to some fixed rule, then we require that the relative frequency so calculated should tend to the same limit as it does in the original set.
>
> 3. The fulfilment of the condition (ii) will be described as the Principle of Randomness or the Principle of the Impossibility of a Gambling System. 4. The limiting value of the relative frequency of a given attribute, assumed to be independent of any place selection, will be called 'the probability of that attribute

within the given collective'. Whenever this qualification of the word 'probability' is omitted, this omission should be considered as an abbreviation and the necessity for reference to some collective must be strictly kept in mind.

5. If a sequence of observations fulfils only the first condition (existence of limits of the relative frequencies), but not the second one, then such a limiting value will be called the 'chance' of the occurrence of the particular attribute rather than its 'probability"' (ibid., pp. 29–29).

"Von Mises' limiting frequency definition of probability has attracted some criticism. Jeffreys argues that to precisely determine the limit of the relative frequency of an attribute, one must perform an infinite number of experiments. This, in practice, is not possible. "No probability has ever been assessed in practice, or ever will be, by counting an infinite number of trials or finding the limit of a ratio in an infinite series... A definite value is got on them only by making a hypothesis about what the result would be" [44].

This objection may have substance if von Mises' probability were taken as an abstract, mathematical notion. However, von Mises makes it clear that probability is a real, physical property analogous to mass, specific heat, or electrical resistance. It is also not possible, due to errors of measurement, to exactly determine what the true value of the mass, specific heat, or electrical resistance of a physical object may be. These also may only be truly determined in the limit, measured with infinite experiments. Yet science seems to progress and make good use of values based on these physical measurements and "a hypothesis about what the result would be". Von Mises would maintain his probability is cut from the same cloth.

2.3.3 Dispositional Probability

As we have seen, the Frequentist approach maintains that probability is an objective, physical property that inheres in a group or collective of events or observations. It is a simple emergent property that can only be rationally spoken about at the ontologically higher, aggregate level. Under the frequentist interpretation it makes no sense to talk about the probability of an individual event except as a subset that is subsumed into a larger class By contrast, the Dispositional account of probability, while also maintaining the ontological existence of probability as an 'in-the-world' property, holds that probability attaches to individual events rather than to collectives. The nature of probability in this case is that of a 'dispositional' property, like being soluble or fragile.

The distinction is made here between categorical and dispositional properties. The former indicates properties that currently occur in the object. The latter refers to properties that are conditional on counter-factual restrictions. Carnap's example of *solubility* serves us here (Carnap 1936). The property of "being soluble in water" cannot mean that an object dissolves when placed in water because this is logically true of any object, however insoluble, if it is never placed in water. What we mean by a water-soluble object is that it *would* dissolve if the object *were* placed in water, which is not true of metal, wood and other insoluble objects even if they are never placed in water. Thus the property of *solubility* is dependent on a set of conditions that are not necessarily realized.

In a similar way, according to the dispositional account, probability is the strength of propensity of an event (an assessment of an object's attribute) to produce a particular outcome. This means that the event has a disposition to behave in a certain manner were certain test conditions met. What then, under this account, does it mean to say that on a roll of a die the probability of obtaining a six is 1/6? The dispositional account asserts that *each roll* has a disposition to reveal a six and that the strength of this disposition is 1/6 of the strength of the disposition to reveal some number between 1 and 6 inclusive.

Karl Popper was major proponent of probabilities being of a dispositional nature.[43] Supporting Peircean "tychism"[44], Popper's conjecture that probabilities are dispositional grew out of considerations of problems arising due to the apparent indeterministic nature of quantum mechanics.

> "If we are committed, or at least prepared, to conjecture the reality of forces, and of fields of forces, then there is no reason why we should not conjecture that a die has a definite propensity (or disposition) to fall on one or other of its sides; that this propensity can be changed by loading it; that propensities of this kind may change continuously; and that we may operate with fields of propensities, or of entities which determine propensities. An interpretation of probability on these lines might allow us to give a new physical interpretation to quantum theory – one which differs from the purely statistical interpretation, due to

[43] This is hardly surprising since Popper holds *all* universals to be dispositional.

[44] *Tyche* for Aristotle was that which happens unexpectedly, yet still has purpose. Here it is used interchangeably with indeterminism.

> Born, while agreeing with him that probability statements can be tested only statistically. And this interpretation may, perhaps, be of some little help in our efforts to resolve those grave and challenging difficulties in quantum theory which today seem to imperil the Galilean tradition" [67].

However, Popper's account appears incompatible with a deterministic view of physical events and we are left to ask some important questions. Are dispositionalists then constrained to apply probability solely to tychistic cases? If an event is fully determined, then are the only values that a probability measure can yield zero and one? Any other values are surely an illusion generated by experimental method. Fully determined events might possibly be assigned non-extreme probability values if probabilities are defined, as Sklar states, by the "relative frequency of the outcome that would result from repeated trials of the *kind* of experiment in question. But what would this frequency be? It would be, given the determinism, fixed by the actual distribution over the initial conditions that would have held" [73].

In this section we have performed a review of theories of probability that, while not by extensive, does examine important aspects of probability in relation to information. In particular the work of von Mises, Bayes and Jeffreys which we will employ in consideration of the relationship between information and probability in Section 4.4. None of the theories reviewed are without their problems. However von Mises' frequentist description appears to be the least problematic and most disposed to practical application, so is this account of probability that I will employ in the development of a unified theory of information.

2.4 Signpost

This concludes this chapter's somewhat cursory review of theories of information and probability over the past hundred years or so. It is valuable here to signpost the relationship of this chapter to the rest of the thesis. It is my intention in this book to show how my account of information – a foundational, group theoretic approach – captures the essence of what we mean by physical information but also how it underlies the existing theories of information that we have reviewed in this chapter and in a sense provides a unification. While the combinatorial/probabilistic approaches of Brillouin and Shannon are close enough to be reconciled, the generative approach of Kolmogorov, Solomonoff and Chaitin is so different from the Brillouin-Shannon approaches as to appear as an isolated field. Indeed people often speak of 'Kolmogorov Information' as if it were a separate class of information. It is my intention to show how these different approaches can be unified by examining what underlies them.

In the next chapter I will turn to the development of the description of the role of asymmetry in information starting with the crucial concept of distinguishability. Then, Chapter 4, I will use the notion of distinguishability to develop my account of the Asymmetry Principle of Information and show how it relates to each of the topics we have discussed in this current chapter.

3

Information and Distinguishability

3.1 Distinguishability

Distinguishability – the extrinsic quality of an object which permits one to say that it is one specific entity and not another or that the object is in one particular state and not another – lies at the heart of a foundational account of information. For if we cannot differentiate between two or more entities or states it is inconceivable that the object can produce any reduction of indeterminacy. More directly, if one cannot, in principle identify properties which determine an object's state then the object is incapable of carrying information.

Much work on distinguishability arose as a result of studies in identity theory, particularly those undertaken by Leibniz. Directed by a monadological account of metaphysics, Leibniz maintained that no two distinct entities in the universe were indiscernible.[1]

> "There is no such thing as two individuals indiscernible from each other" [53]. Also "It is necessary, indeed, that each monad be different from every other. For there are never in nature two beings which are exactly alike and in which it is not possible to find an internal difference, or one founded upon an intrinsic quality (*dénomination*)" [52]

Leibniz here intends that each two entities will always differ in some properties, even if those properties rest at such a level of subtlety that

[1] I will use, as Leibniz did, distinguishability and discernability synomously. Argument might conceivably be made for a case wherein discerning is a mind-mediated activity while distinguishing has no such restriction requiring only a differential sensing capacity. This would serve no useful purpose here as discerning entails at least distinguishing and, as shall be seen shortly, we are interested in a most general, objective theory of information.

it eludes all natural observers. Two entities, drops of milk say, may appear to be identical when observed with the naked eye, however "viewed with a microscope, [the drops] will appear distinguishable from each other" (Alexander, loc. cit.). This is Leibniz's *Principle of Indiscernibles.* There will always exist, at some level, qualities of an entity which differentiate it from every other entity in the universe. While I do not intend to pursue the veracity of this principle[2], I do wish to utilise Leibniz's notion that entities have properties existent on manifold levels, all with varying degrees of accessibility relative to an observer which allow them to be distinguished from other entities.

It is at this point I wish to differentiate two levels of discernability: *discernability in principle* and *discernability in practice*. Leibniz was primarily concerned with discernability in principle – at some level, physical differences exist between all entities such that no two are identical. Such properties are those which could be apprehended by a trans-natural being with arbitrarily acute perception. Such a being could apprehend *all* properties of an entity (even if they were uncountable in number). The capacity to distinguish those properties which can be perceived by a particular real natural system[3] with finite and specific powers of apprehension I term *discernability in practice.* Discernability in practice is always "with respect to" a perceiving (or discerning) system and its limits of perception. Leibniz I will move early here to quash any tendency to suggest subjectivism. The property differences that facilitate distinguishability are physical and objective. They exist independently of the perceiving system. Just as independently as those additional properties that the system may not perceive. The system defines the closure of the set of properties that can be perceived in practice with respect to that system, but not the elements themselves.

In order to illustrate the idea of levels of distinction, consider the following scenario. Imagine that, on June 20, 1908, two years after Nernst proposed his theorem, in the safe of gem trader M. Stéphane Rolland's office on Avenue de l'Opéra in Paris, there are two labelled black velvet bags each containing a single cut diamond. They belong individually to M. Gustav Hofmann and M. Charles Bloit and have been given into M. Rolland's care awaiting the arrival of the diamond assessor M. Van Der Hooft who visits on the 25th of each month from

[2] I shall also leave aside more modern considerations of the semantics of identity (Frege et. al) because I wish to deal at this point solely with an objective entity and leave aside epistemological and semantic considerations.

[3] Here I use the term 'system' as a finite dynamic interacting entity. It is intended to be the natural perceiver corresponding to the 'trans-natural being' of the previous sentence.

Amsterdam for appraisals. M. Van Der Hooft is renown throughout Europe as the most experienced diamond assessor in the business. On the 25th M. Van Der Hooft examines the two diamonds and is astounded to find how remarkably similar they are. They have obviously been cut by the same person in the exactly same manner but more than that, they are of identical size and weight and both are flawless, even under ten-times magnification. After hours of exhaustive examination with his best instruments which stretches his expertise, he remarkably pronounces them both to be 'identical': weight of 2.31 carats each and to be worth 9460 Fr each.

As it turns out, by the most improbable coincidence, the structure of the two diamonds was identical. Diamonds are typically octahedral carbon crystals with covalent bonds.[4] In the case of the Hofmann/Bloit diamonds, the number of atoms was the same in each diamond. For each carbon atom in one diamond, there was another in a corresponding position in the other diamond. Being perfect, both diamonds had no dislocations. In short there was no technique available to M. Van Der Hooft which would permit him to distinguish between the diamonds.

However, during the course of the rigorous examination, the provenance of each diamond is confused: M. Rolland and M. Van Der Hooft are unsure which diamond came from which bag. After some discussion they reason that since the diamonds are identical, it doesn't matter to which bag the diamonds are returned. It is at this point that the mistake is made and the diamonds are returned to the wrong bags. Although no difference between the diamonds could be distinguished by M. Van Der Hooft, there in fact exists a dissimilarity which may possibly make one of the diamonds more valuable than the other.

Carbon naturally occurs on Earth primarily as either of the two stable isotopes Carbon 12 (C_{12}) or Carbon 13 (C_{13})[5] with a C_{12}: C_{13} ratio of about 98.9 : 1.1. Natural diamonds usually occur with this ratio and this was the ratio for M. Hofmans's diamond. M. Bloit's diamond, on the other hand, was pure C_{12} with no C_{13} isotope present at all. Thus the C_{12}: C_{13} ratio was 100.0 : 0.0.

Isotopes are chemically indistinguishable from each other, detectable only by such methods as NMR spectroscopy and X-ray photoelectron spectroscopy. M. Van Der Hooft had no capacity of discovering this

[4] A crystal is an regularly repeating three dimensional structure held together by chemical bonds (either ionic or covalent).

[5] C_{13} is a carbon isotope with one more neutron than C_{12}..

isotopic difference.[6] Indeed, he almost certainly couldn't know that such a difference was possible as it wasn't until 1910 that the existence of isotopes was determined by Fredrick Soddy.[7] The presence of the extra neutrons in the C_{13} nuclei means that these atoms are capable of creating bonds with other atoms that are stronger than those of C_{12}.[8] This implies that M. Hofmann's diamond would have been slightly harder than that of M. Bloit and thus potentially more valuable. (This fact might have been determined by M. Van Der Hooft by means of destructive testing, however, such tests are precluded by his role as an assessor.)

We can see that while the two diamonds were distinguishable in principle, they were not distinguishable in practice with respect to M. Van Der Hooft. Of course it is possible that a different perceiving system could distinguish between C_{12} and C_{13} – a researcher with a magnetic resonance device, for example. But we do not require anything as sophisticated as modern researchers to distinguish between isotopes. Plants have recently been shown to have the ability to discriminate the two stable Carbon isotopes in their photosynthesis process [12]. A distinguishing system doesn't even need to be an animate system. At the molecular level, chemical processes distinguish reactants on the basis of differences in molecular properties. A particularly insightful example is that of stereo-selectivity in enzyme mediated reactions. Enzymes are large organic molecules, usually proteins[9], which catalyse certain reactions between two or more substrates. This is achieved by binding to the reacting substrates, orienting them in a manner conducive to the reaction. Different groups of enzymes vary in the degree of specificity with which they catalyse reactions. A great many enzymes are highly specific; particularly the dehydrogenases, the kinases and the synthetases [30]. Mostly these enzymes are highly specific for both reactants for bimolecular reactions or all three reactants in the case of trimolecular reactions. A few enzymes are specific for one reactant and not another such as alcohol dehydrogenase which is specific for NAD but the other reactant may be any alcohol [30].

---

[6] Though, perhaps it is possible that if Van Der Hooft had a sufficiently accurate balance, the mass difference of the additional neutrons may have been detected.

[7] The existence of isotopes was recognized by independent work of Soddy, T. W. Richards and J. J. Thomson from 1907 to 1913 though Soddy is generally credited with the discovery (See Aston, 1942).

[8] It is interesting to note that General Electric, the creators of the first artificial diamond in 1955, made a pure C_{13} diamond 1991 which has been found to be substantially harder than C_{12} diamonds.

[9] Though messenger RNA is also known to possess catalytic ability.

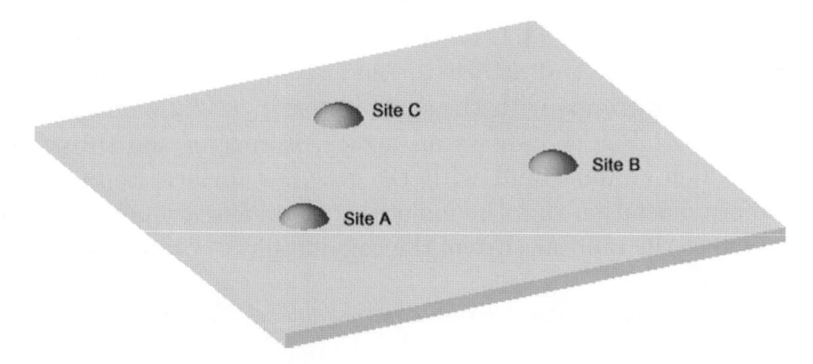

Fig. 3.1. Enzyme surface with three-point attachment

Beyond being chemically specific for reactants, enzymes are also sterically specific, meaning that they will discriminate between isomeric reactants with asymmetric centres.

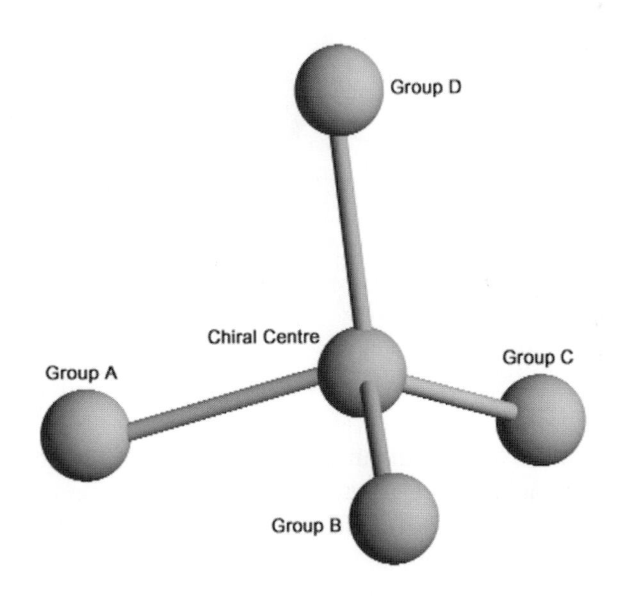

Fig. 3.2. Tetrahedral Substrate

Most substrates will form at least three bonds to attach to an enzyme (three-point attachment) [60], thus if a substrate has a tetrahedral structure, such as a carbon atom as a chiral centre, an enzyme will typically bind with one optical isomer and not the other. Fig. 3.1 shows a stylised depiction of three-point attachment sites on the surface of a large enzyme, while Fig. 3.2 shows a schematic of a reactant which has the ability to orient and attach to the enzyme. In the example, sites have affinity for groups with the same letter.

By contrast, Fig. 3.3 shows an optical isomer of the substrate in Fig. 3.2 which cannot bind with the enzyme surface as there is no rotation or translation that the molecule can undergo which will permit it to attach to the binding sites of the enzyme short of breaking covalent bonds.

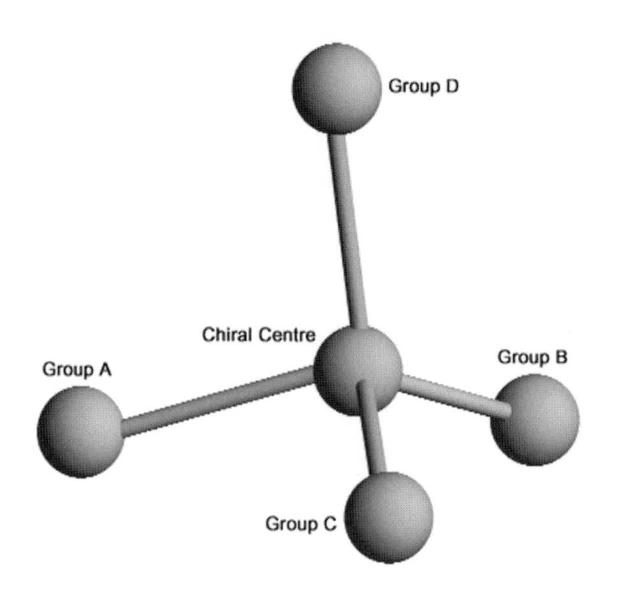

Fig. 3.3. Optical Isomer

This enzymatic distinguishing is crucial for biological processes, since all amino acids have a carbon chiral centre (known as the α-carbon) and thus all have optical isomers (denoted L-α and D-α). Yet with some rare exceptions[10], enzymes act almost exclusively on L-isomers of amino acids. Indeed, only L-α-amino acids form proteins.[11]

Returning to the discourse on distinguishability in practice, what is important to note in this example is that, while the groups on the

[10] Kidney D-amino acid oxidase for example.

[11] Why this is the case is open to speculation. D-α amino acids do occur in nature, but perhaps some primordial, minor perturbation caused the L form to dominate and, combined with feedback via enzymatic stereo-selectivity, evolutionary 'lock-in' occurred.

substrate may be chemically identical, they may become distinct with respect to the enzyme. For example, if in Fig. 3.2 site A represented a site with affinity for the carboxylic acid group (–COOH) and, in Fig. 3.2 both groups A and D were carboxylic acid, the enzyme would be able to distinguish between the two groups because of their positions relative to the other two groups and the chiral centre: there is no rotation which would permit group D to bind to site A *and* groups B and C to bind to sites B and C respectively. It is at a systemic level that distinction between components may be made. Other chemical reactions may allow optical isomers to react at equal rates; however, these isomers are distinguishable in practice *with respect to* certain enzymes.

We may at this point formally state the notion of distinguishability in practice:

Definition 1. *Let Θ_P denote the set of all properties of an entity P and $\Phi_{P|X} \subseteq \Theta_P$ denote the set of those properties of P that can be discerned by a system X. Similarly, let Θ_Q denote the set of all properties of an entity Q and $\Phi_{Q|X} \subseteq \Theta_Q$ denote the set of those properties of entity Q that can be discerned by X. Then X can* ***distinguish*** *between entities P and Q iff $\Phi_{P|X} \neq \Phi_{Q|X}$.*

Note that if it were the case that $\Theta_P \equiv \Theta_Q$, then P and Q are indistinguishable in principle, a possibility that Leibniz would reject. However, I concern myself here only with distinguishability in practice, that is with reference to a distinguishing system, in this case X.

Thus far in this section, I have considered examples of distinguishing between entities: discerning that they are indeed different. What is often of interest is distinguishing between different states of the same entity: discerning change in an entity. By *state* I mean a set of values of properties whether those values are capable of being different or not. The *state space* is the set of all possible states. If the state space is well defined for a particular entity then we can treat distinguishability in practice of multiple states of an entity in a manner similar to distinguishability in practice of entities, that is, by establishing whether a perceiving system is capable differentiating two or more states. Consider the following example. Gretel bought an expensive stylised wall clock from an on-line Austrian Bauhaus design studio. When it arrived, some time later, she removed it from the box and packaging and noticed the following features:

- The clock was perfectly circular
- There were no markings at all on the face

- The hour and the minute hands were of equal length and width and of identical material and colour
- It made absolutely no noise,

everything one would expect from an Austrian Bauhaus wall clock.

Fig. 3.4. Clock at time t_0

She placed the clock on the kitchen table and left the room. When she returned some time later, she found her flat mate Hansel holding the clock, rotating it this way and that, trying to determine which way up it should be mounted on the wall. "I can't work it out." said Hansel to Gretel as he replaced the clock on the table and left the room. (Fig. 3.4) Gretel went to the table and looked at the clock. The thought then struck her that, since the clock made no sound, it might not be working. She remembered the position of the hands when she had first laid the clock on the table at time t_0 and was now (time t_1) looking an apparently different configuration (Fig. 3.5).

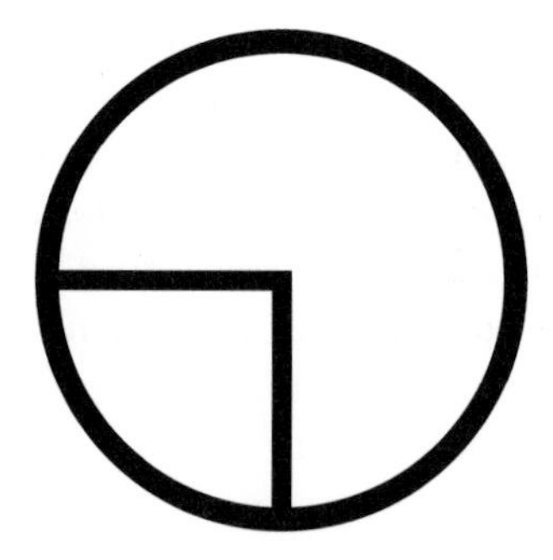

Fig. 3.5. Clock at time t_1

Gretel estimated that the angle between the hands at both times was the same at about 90° and estimated that the time that had passed between t_0 and t_1 was around 30 minutes. After a moments reflection she correctly concluded that, given such information, it was not possible to assert whether the clock was working or not.

What Gretel had determined was that given her capacity to apprehend the properties of the clock and given the rotational symmetry of the clock, it was not possible to distinguish between the two states of the clock. The states of the clock here are the positions of the hands relative to each other and the clock face. The clock hands may have remained motionless and the whole clock rotated 180° the clock hands may have moved as designed over almost 33 minutes and the whole clock rotated some 73.5°. All that may be concluded for certain is that the whole clock has been rotated by some amount.[12] Thus the states of the clock at t_0 and t_1 are indistinguishable in practice relative to Gretel.

With these considerations as a starting point we can define distinguishable states in a manner similar to distinguishing entities,:

Definition 2. *Let* Θ_P*denote the set of all states of an entity P (the state space of P) and* $\Phi_{P_{t_0}|X} \subseteq \Theta_P$ *denote the set of those states of P that can be discerned by a system X at time* t_0. *Let* $\Phi_{P_{t_1}|X} \subseteq \Theta_P$ *denote the set of those states of P that can be discerned by X at time* t_1. *Then X can* ***distinguish*** *between states at* t_0*and* t_1 *iff* $\Phi_{P_{t_0}|X} \neq \Phi_{P_{t_1}|X}$.

As a final example of distinguishability, and to lead us into considerations of information, imagine the transmission of a ten binary digit signal which serves as an error state indicator on a deep-space probe. The signal is transmitted to earth receiving stations at a certain frequency in the event of equipment failure on board the probe. The ten-bit signal is capable of identifying 1064 different, pre-determined equipment states. Long before launch, the design decision was made that the signal should be continuously repeated to allow for the possibility of some signal loss, interference or earth reception problems (eg, no-one listening at a particular moment). This, however, had some unfortunate consequences. Consider the signal that was received by a deep space tracking station:

[12] This is assuming that Gretel is looking at the clock from the same place, thus imposing an arbitrary external reference point. It might also be possible that Gretel may have been grossly mistaken regarding the time between t_0 and t_1, in which she cannot say anything, even about the clock being rotated.

...110001010111000101011100010101110001010111 00...

Interpretation of the signal is difficult, since the encoded number is not uniquely distinguishable. The number could be 1100010101, decimal 789, which indicates a low-power state of a backup battery or it could be 1000101011, decimal 555, a receiving antenna failure. Or it could be 87 or 174 or 348, or five other numbers. There are ten possible numbers, only one of which corresponds to an actual equipment failure on the probe. Because of the repetition, these ten numbers are in distinguishable from each other. The problem might have been solved if some sort of end-of-string indicator had been used.

3.2 Information: A Foundational Approach

What are the fundamental requirements for an entity to carry information? As noted in the previous section, for any system to reduce indeterminacy concerning an entity or other system, it must in practice be able to distinguish properties of that entity or other system. Specifically, it must firstly be able to distinguish between the object and the environment and it must secondly be able to distinguish which of the possible states the object is in. A system that is capable of making such distinctions I will call an *Information Gathering and Using System* (IGUS) after Zurek [89].

For an object to carry information beyond its existential bit[13] that can be detected by an IGUS, it must have more than one possible state that can be distinguished in practice with respect to the IGUS. If we consider Gretel's Austrian Bauhaus wall clock as an object and Gretel as an IGUS, for the clock to carry information perceivable by Gretel (that is, for it to be an information object with respect to Gretel) then it must be capable of being in multiple states and Gretel must be able to distinguish between these states in practice. For Gretel to actually tell the time from the clock, she must be able to identify unique hand arrangements relative to the clock face.

An information object may carry many more distinguishing features than can be apprehended by a particular IGUS. Gretel's clock, for example, may have had hands which differed in colour wavelength by a few Angstroms, an amount imperceptible by Gretel but that would have enabled her to identify more states had she been able to observe it. This is again the 'in principle'/ 'in practice' distinction.

[13] That is, that binary distinction that it is something rather than nothing.

It would be nice to evaluate the *maximal information* of an entity, that is, the information that is based on all the distinguishable-in-principle properties. This would give us an absolute limit to the amount of information that the object is capable of carrying. However, there is a difficulty associated with this. It is the problem of identifying these properties, for if we are to identify them, they must be physically observable and hence distinguishable in practice. We know that not all properties are observable in arbitrarily small detail.

This forces us to consider an alternative to the maximal information approach: an evaluation of information content with respect to a particular IGUS. The set of states that can be distinguished in practice is, in a conceptual sense, generated by passive filtering due to the IGUS's inherent inability to apprehend the complementary set of states. The information that the object can carry in this 'co-system' is dependent on the IGUS.[14] I again stress that this is not a subjective definition. The object's distinguishing properties exist independently of the IGUS. The information here exists as a *relational ontology*. It can be represented by the schematic model shown below in Fig. 3.6.

Fig. 3.6. Information Model

The information question then becomes: Given a specified set of filters of a particular IGUS, what is the informatic capacity of an object? To answer this we must look at the number of ways the object can be distinguished with respect to the IGUS. The more states an object can possibly be in, the more information it can carry. But it is not sufficient to combinatorially count off the number of possible states in which an object may exist with respect to a certain property defined state space. We have already seen that, although an object may

[14] It should be noted here that the transmission of information is assumed to be conducted perfectly with no "line losses". In real systems this is not the case. Loss of information in the channel between the object and the IGUS can occur due a number of factors including external interference, noise and the nature of the channel itself (e.g., a person who can see colours cannot see them in the dark). The model could be modified to incorporate these losses by including external filters between the object and the IGUS.

be physically capable of being in two physically different states, these may be indistinguishable to a certain IGUS. Gretel was incapable of distinguishing between the clock at t_0 and t_1, thus the observations of the clock cannot be held to have reduced indeterminacy. What needs to be accounted for is the possibility of states which appear to be identical due to the relationship of the object and the IGUS.[15,16] The physics of the information object and the set of filters pose constraints on which certain transformations are possible. These transformations are the transitions which govern the movement of the object between states *which can be observed by the IGUS.*

[15] This is the generalization of the degeneracy principle in statistical mechanics. This will be developed later.

[16] It is somewhat similar to Rosen's Equivalence Principle [68] though Rosen's principle was nomic in nature.

4

Information and Symmetry

4.1 Symmetry

The concept of symmetry is an ancient one and has made a great contribution to the formation of our intuitions. This is rightly so, given its relationship with information, which will become evident in the course of this chapter. Etymologically, *symmetry* originates from the Greek *συμμετρον* (sym metron – "alike measure") and in its most general form, symmetry denotes a balance of form; a distribution of parts in like-relations to form an integrated whole. The physical world abounds with symmetry. D'arcy Wentworth Thompson's classic *On Growth and Form* contains a great many examples of natural symmetries, from cell structures to radiolarians to animal horns, "symmetry is highly characteristic of organic forms and is rarely absent in living things"[80].

Symmetry can take many forms. The most accessible are the three dimensional geometric symmetries: reflection, rotation and translation. These are found in abundance in the physical world. Bilateral symmetry, the symmetry of left and right, obvious in the human form, is an instance of reflection. The repeating octagonal structure in M. Bloit's diamond is an example of finite translational symmetry. And a falling milk drop, spherical to minimise surface energy, is rotationally symmetrical around any axis.

However, symmetry is not confined to the purely natural world. Symmetry finds residence in art, architecture and design. We find it appealing. Why? Is this because of its presence in biology as Thompson has noted? Hermann Weyl considers this question:

> "One may ask whether the aesthetic value of symmetry depends on its vital value: Did the artist discover the symmetry with which nature according to some inherent law has endowed

its creatures, and then copied and perfected what nature presented but in imperfect realizations; or has the aesthetic value of symmetry an independent source? I am inclined to think with Plato that the mathematical idea is the common origin of both: the mathematical laws governing nature are the origin of symmetry, in the intuitive realization of the idea in the creative artist's mind its origin in art; although I am ready to admit that in the arts the fact of bilateral symmetry of the human body in its outward appearance has acted as an additional stimulus" [87].

Symmetry exists beyond just the three spatial symmetries noted above. Reversible processes such as the frictionless dynamics of spheres with elastic collisions are temporally symmetric. Mathematical abstractions such as periodic (e.g. trigonometric) functions are symmetric. So Weyl's suggestion that mathematical laws are the origin of symmetry has a great deal of appeal. If symmetry is generated by natural mathematical laws, then it would be possible to construct a formalization of symmetry, a calculus of form. However, before proceeding with an examination of how such a formalization of symmetry would be performed, it is perhaps appropriate here to take pause and consider the following questions: If symmetry is governed by mathematical laws, why should it occur where it does in such abundance and be 'highly characteristic'?; and "How are these laws given physical realizations?" The answer to these questions is complex and involves the consideration of equilibria.

Consider M. Bloit's pure C_{12} diamond. The reason that the octagonal structure was adopted and held by the carbon atoms is that, given the environmental constraints (temperature, pressure etc) and the attracting and repelling atomic forces, the octagonal structure was the optimal structure for minimising the energy equation of the carbon atoms. Subject to different conditions, the carbon atoms would adopt a different structure. Under difference physical condition carbon takes another crystalline form, graphite, which has a hexagonal crystal structure as a minimising solution to the energy equation.[1]

But, one could point out, this is another symmetric structure. Why can't an amorphous structure be a minimal solution? Having atoms separated at equal, regular distances averages out the internal stresses, so each of the C–C bonds shares the workload. If the structure was amorphous, then there would be certain areas in which atoms were packed very close together, forming pockets of intense stress. The compressed

[1] Elements that take manifest in different physical structures are known as allotropes. Other Carbon Allotropes include Fullerene C_{60} and C_{70}.

atoms would have to move and push other adjacent atoms in order to reduce inter-atomic repulsive forces. The displaced atoms would themselves have to move to relieve the increased stress. This would continue en masse until the sparse regions and compact regions have been averaged out and, in a Pareto-like manner, no net gain would be attained by any more movement, so the energy equation is at a minimum. This is an example of the principle of least action of work. The process in this argument is generalised in the following quote by Ernst Mach:

> "In every symmetrical system every deformation that tends to destroy the symmetry is complemented by an equal and opposite deformation that tend to restore it. In each deformation, positive and negative work is done. One condition, therefore, though not an absolutely sufficient one, that a maximum or minimum work corresponds to the form of equilibrium, is thus supplied by symmetry; there is no reason, therefore, to be astonished that the forms of equilibrium are often symmetrical and regular" [56].

Mach's observation here is significant in that it proposes a relationship between symmetry and work. Later we will examine the capacity of information to generate work. This is directly related to the symmetry-work relationship noted by Mach. For now we will note that a formal account of symmetry must account for the invariance under change that gives symmetrical objects their special qualities. This is the starting point for the next section.

4.2 Symmetry and Group Theory

In order to develop a formal definition of symmetry, we will first establish some supporting definitions. A one-to-one mapping M associates every point in space p with another point p'. In the trivial case that M maps p on to itself (that is $p = p'$), then M is called an *identity mapping*. The mapping of p' back on to p is called an *inverse mapping*. A *transformation*, then, is the pair of one-to-one mappings M and its inverse. Any transformation that preserves the structure of space[2] is said to be an *automorphism*. Consider two mappings, M, which maps p to p', and N, which maps p' to p'' and their respective inverses M^{-1} and N^{-1}. Denote MN as the consecutive application of mappings M then N. Now it should be obvious that MN maps p to p'' and that if

[2] That is, relationships of congruency are maintained.

the transformation M is an automorphism and the transformation N is an automorphism then MN is also automorphism.[3] The inverse of MN is $(MN)^{-1} = N^{-1}M^{-1}$. The consecutive application of the transformation is sometimes called 'multiplication' and is not necessarily commutative.[4]

Definition 3. *Define G to be the set of all automorphisms. G is said to be a group if the following conditions are satisfied:*

1. *Multiplication is associative, that is* $(MN)P = M(NP)$
2. *G contains the identity mapping transformation (designated* e*)*
3. *Each automorphism M has an inverse* M^{-1} *which is also a member of G*

It should also be noted that the multiple of any elements of G is also a member if G.

Just as numbers measure size, groups measure symmetry. General Group Theory grew out of early work on specific permutation groups by LaGrange, Ruffini, Cauchy and Galois in the late 18th and early 19th century. Felix Klein applied Group Theory to geometry in the second half of the 18th century. To see how groups apply to symmetry, consider an automorphism M which maps points p to p'. Now consider a spatial figure in that space prior to application of the transformation. A spatial figure may be thought of as a set of points, S, with a defined relationship to one another. If after the application of M, the figure is indistinguishable from its pre-transformation state, M is said to be a *symmetry* of S. A group G consisting of all automorphisms M which leave S invariant exactly describes the entire symmetry of S. The order of the group denotes the number of symmetries.

In physical systems, 'possible transformations' can be used to form a group to determine the symmetry of an object. A possible transformation means that it is physically possible to perform the change to the system, though it doesn't actually have to be performed. If a transformation is possible and its application would result in an arrangement that is indistinguishable from the original, the transformation is known as a symmetry. If a transformation is possible and its application would result in an arrangement that is distinguishable from the original, the transformation is known as an asymmetry.[5] As we have noted in the

---

[3] Strictly the transformation formed by M and M^{-1}.

[4] That is, MN is not necessarily equal to NM.

[5] Symmetry theorists also talk of a case of dissymmetry in which the transformation type may yield either distinguishable or indistinguishable instances.

previous section, distinguishability is the foundation of informatic capacity, thus we can conclude that it is the set of asymmetries which generate the informatic capacity.

To clarify the use of Group Theory to describe physical symmetries, let us consider the geometry of a tetrahedral carbon compound such as that presented in our considerations of optical isomers. For this exercise, assume the chemical groups A,B,C and D that were shown in Fig. 3.2 to be the same chemical group (e.g. hydrogen, thus forming methane) and to occupy the apexes of a stylised figure (see Fig. 4.1). Thus we can represent the molecule as a simple tetrahedron. By inspection, we can determine the transformations that can be applied to the tetrahedron.

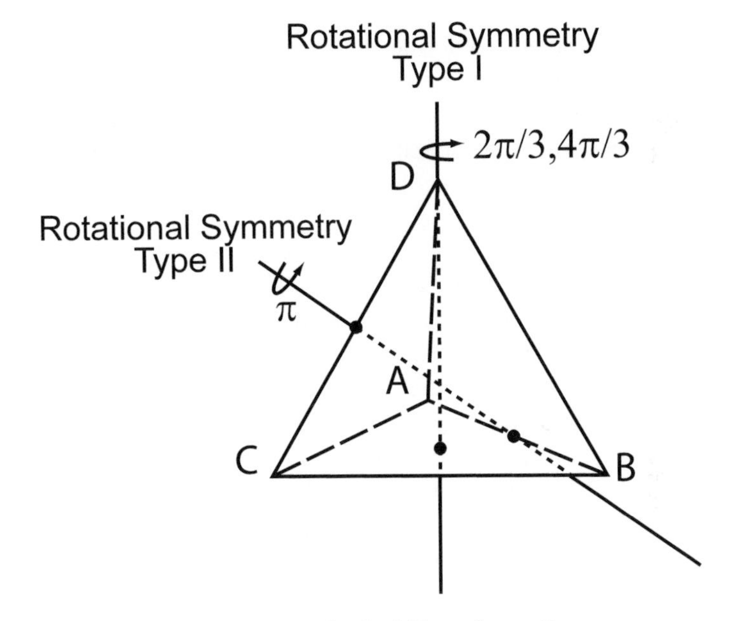

Fig. 4.1. Tetrahedral Transformations

Since reflective transformations would involve bond breaking, these are excluded as they are physically impossible. That leaves us with rotations. There are two kinds of rotational symmetries. The first is a rotation about an axis which runs through an apex and the centroid of the opposite face. Rotation by either a third or two-thirds of a turn will leave the tetrahedron in a similar position. This kind of rotation I shall call type I. The second kind, type II, is rotation about an axis which runs from the middle of one edge, through the tetrahedron's centroid, to the middle of another edge. Half a turn will result in a similar configuration.

Each of these rotations generates a transformation: the rotation and its inverse.[6] To construct the transformation group it is necessary to count the total number of transformations. Designate type I rotational transformations as r^1 and type II as r^2. There are 4 faces, so there are 2×4 r^1 transforms which we will label $r_1^1, r_2^1, r_3^1, r_4^1, r_5^1, r_6^1, r_7^1, r_8^1$. Further there are 6 edges, hence 3 r^2 transformations: r_1^2, r_2^2, r_3^2. The final transformation to be included is the identity transformation, e, which entails no action or application of a rotation around any axis through 2π. So the set of rotational transformations is:

$$\{e,\ r_1^1, r_2^1, r_3^1, r_4^1, r_5^1, r_6^1, r_7^1, r_8^1, r_1^2, r_2^2, r_3^2\}.$$

But this is not sufficient to define the group. The set of rotations has an internal structure in that they have algebraic features when combined. For example, $r_1^1\ r_1^1 r_1^1 = e$ and $r_1^2 r_1^1 \neq r_1^1 r_1^2$. The combination or 'multiplication' method must be specified to define the group. Designate this as m. Hence the group is specified as:

$$G = \{e,\ r_1^1, r_2^1, r_3^1, r_4^1, r_5^1, r_6^1, r_7^1, r_8^1, r_1^2, r_2^2, r_3^2\}, m.$$

Every multiplication of the elements of G results in a transformation which is also a member of G. Note that in this example we have set chemicals A, B, C, D to be identical. It is this that makes the symmetries under rotational transforms. If, for example, we nominated A, B, and C to be hydrogen and set D to be chlorine, thus forming chloromethane, then the number of symmetries would be reduced, since some, but not all, rotations would produce results distinguishable (by the relevant IGUS – here the enzyme) from the original configurations. Changing another hydrogen to chlorine to form dichloromethane produces new symmetries. In doing this, the group G remains unchanged, of course, since it is just description of the spatial geometric algebra – of what rotations are possible. It provides no guarantee of symmetry. What we require is a method of measuring only those transformations that produce results which are indistinguishable from the initial states. To do this, let us further develop the notion of the application of a transformation to a set.

Let G be a group and S a set and define an *action* of G on S to be a homomorphism, which for each element g of G gives us an arrangement of the points in S. The transformations in G act on the points in set

[6] The inverse is just rotation of the same magnitude in the opposite direction. Note that the type II symmetries are their own inverses.

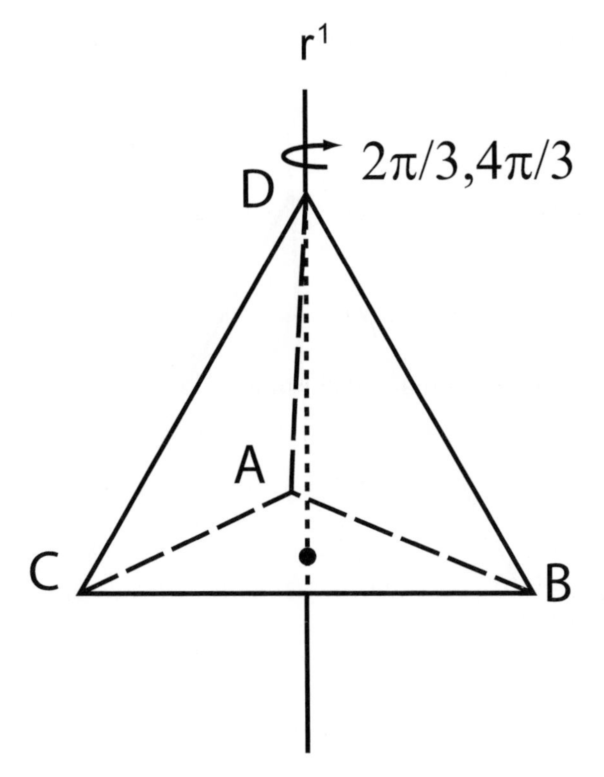

Fig. 4.2. Single Rotation

X and permute them in a manner that is consistent with the algebraic structure of G. For a given action of G on S and a point $s \in S$, an *orbit* is defined as the set of images of $g(s)$ as g varies through G. By way of illustration, consider the tetrahedron example. Let S be composed of the edges of the tetrahedron = {CD,BD,CB,AC,AB,AD}. To simplify matters, consider only one r^1 rotation around a single axis (see Fig. 4.2). Thus $G = \{e, \mathrm{r}^1_{2\pi/3}, \mathrm{r}^1_{4\pi/3}\}, m$. The action of G on each of the elements of S produces two sets of images, illustrated in Fig. 4.3 by solid and dotted lines.

These two distinct orbits partition the set S in such a way that for two sides to be genuinely different they cannot both lie in the same orbit. Under the action of the group generated solely by rotation about a single axis as shown above, CD, AD, DB are indistinguishable from one another but are distinct from sides CA, AB, CB. Moreover we can say that the number of distinguishably different permutations is given by the number of orbits. Group theory provides us with a theorem for determining the number of distinct orbits. The Burnside Lemma (proof

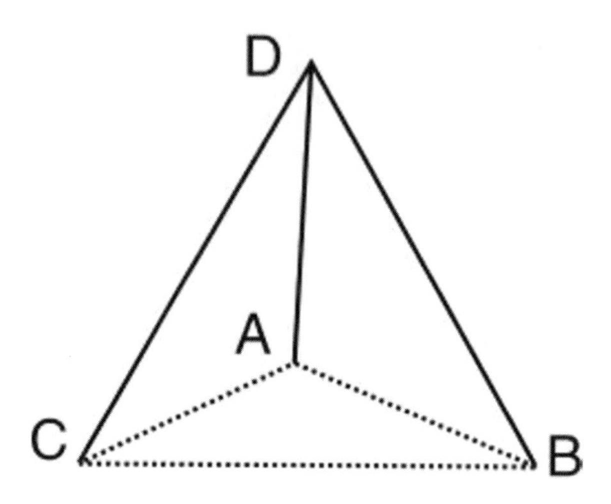

Fig. 4.3. Tetrahedral Orbits

provided in Appendix A) states that the number of orbits, O, is given by the average number of elements in set S which are 'fixed' – that is left unchanged – by a transformation g in G.Thus:

$$\mathrm{O} = \frac{1}{|G|} \sum_{g \in G} |S^g|$$

where $|G|$ is the order of the group G and $|S^g|$ is the order of the subset of points in S fixed by g. As an illustration of determining the number of distinguishable permutations of an object by count orbits, consider the problem of painting each of the faces of a cube with one of three colours; red, white and blue as presented in Neumann et al [61]. Given the rotational transformations that could be applied to the cube, how many distinguishably different ways are there of painting the cube? There are 3 types of rotational symmetry associated with a cube: Type I, rotation about an axis which runs through the centre of a pair of opposite faces; Type II, rotation about an axis which runs through diagonally opposite corners; and Type III, rotation about an axis which runs through the midpoints of a pair of diagonally opposite edges (see Fig. 4.4).

There are six Type III axes each capable of a rotation transformation of π. Rotation of π about one of these Type III axes interchanges the

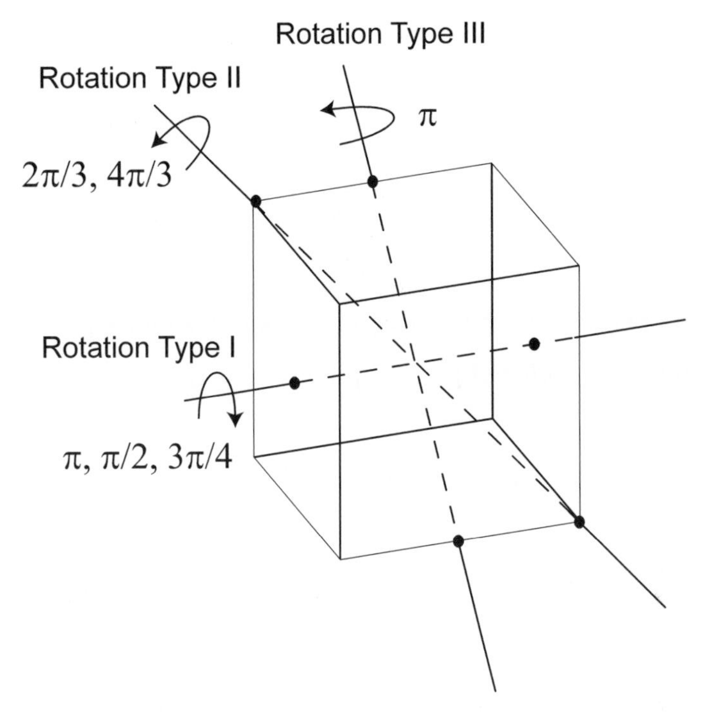

Fig. 4.4. Axes of Symmetry for a Cube

following faces: top with north, bottom with south and east with west. Thus if the top face and the north face have the same colour (red, white or blue), the bottom face and the south face have the same colour, and the east and west faces have the same colour, then the rotated cube will be indistinguishable from the original cube. Thus there are 3^3 members of S which are fixed by this transformation. The total number of s which are fixed by Type III rotations then is $6 \times 3^3 = 162$. This calculation is repeated for all transformations yielding a total number of s fixed to be 1368. There are a total of 24 transformations in this group, hence the number of orbits is $1368/24 = 57$. That is number of distinguishable cubes, or to put it another way, there are 57 complexions of the cube.

4.2.1 Subgroups and Special Groups

We have seen that groups are composed of symmetries and that some of the symmetries are related to each other by virtue of them being the same "type". In Fig. 4.1 we noted that there were two types of rotational symmetry: type I rotational transformations which we labelled as r^1 and type II which we labelled as r^2. The full group of rotational transformation is defined as:

$$G = \{e,\ r_1^1, r_2^1, r_3^1, r_4^1, r_5^1, r_6^1, r_7^1, r_8^1, r_1^2, r_2^2, r_3^2\}, m.$$

We can take a subset of G as follows: $H = \{e, r_1^2, r_2^2, r_3^2\}$ with the same multiplier m as G and see that we have another group. The inverses are present: $(r_x^2)^{-1} = r_x^2$, the identity element is present and the product of any two symmetries is a member of the set. Here we say that H is a subgroup of G under the following definition of subgroup:

Definition 4. *A **subgroup** of a group G is a subset of G which itself forms a group under the multiplication of G.*

In our introduction to Group Theory we have concentrated on groups based on geometric shapes. Group certainly aren't limited to applications involving physical geometry but geometric groups form an import subclass. For example, the symmetry group describing a perfect sphere, accounting for rotation through all angles about any axis through its centre and reflection in planes along these axes, is termed the **orthogonal group** and is represented by O(3). The group consisting of just the rotations is called the **special orthogonal group**, SO(3). Similar symmetry groups describing a circle in two dimension are denoted by O(2) and SO(2). Note that O(2) is a subgroup of O(3) and SO(2) is a subgroup of SO(3).

Groups which describe symmetries of n-agonal plates (that is a thin-plate three dimensional plate with n equal length sides) constitute a family of symmetry groups known as **dihedral groups** and are denoted by D_n. If we have a triangular plate (n = 3), we denote:

$$\mathrm{D}_3 = \{e,\ r,\ rr,\ s,\ rs,\ rrs\ \}$$

where r = rotation $2\pi/3$ around the axis that runs orthogonal to the plate's face through the centroid and s = rotation of π around one of the three axes that run through the plate, parallel to the face at a depth of half of the plate's thickness emerging from the plate at an apex and at the midpoint of the opposite side.

The bulk of our discussion will not be concerned with the application of subgroups. However, as we will see in the Chapter 5 subgroups are crucially important tools in the analysis of symmetry breaking processes and thus, information generation.

4.2.2 Group Theory and Information

The 'Burnside' methodology described in Section 4.2 eliminates duplication produced by symmetries, in effect determining the asymmetries of the object. In determining the transformation group G for an IGUS/object system, it is important to include **all** those possible transformations that the IGUS is capable of detecting (i.e. all those distinguishable in practice. If redundancy exists, that is, if there are symmetries, the orbit counting will factor them out, leaving us with only the number of distinguishable states. If all possible (in practice) transformations are included in the group, then all information that the object is capable of transferring to the IGUS will be accounted for. If any other states of the object exist due to differences in principle, then they can carry no information for the IGUS. In terms of Shannon-type information theory, all the distinguishable states are equally likely, for if they weren't we would have an unaccounted for asymmetry; there would an in-practice distinction that could be made which was not included in the transformation group. We will elaborate on this shortly, but first consider how this methodology applies to our examples. First we consider Gretel's Bauhaus clock. In order to make the calculations simpler, we will impose some assumptions. First assume that the mechanism is such that the minute hand is always pointing at a minute divisor and moves between the divisors infinitely quickly; that is to say, the minute hand can be in one of 60 possible states. Second, assume that the hour hand is always pointing at an hour divisor and moves between the divisors infinitely quickly on the change of the hour; the hour hand can be in one of 12 possible states. In total there are 720 formal states.

Now given the symmetries of this unique clock, it can be shown that by counting the orbits of the group acting on the clock, the total number of distinguishable states is 102 (see Appendix B). Hence, the total information capacity of such a clock is $\log_2 102 = 6.672$ bits. Imagine now that we made the clock somewhat more practical by making the hour and minute hands have different lengths (see Fig. 4.5).

We can demonstrate that this increase in asymmetry increases the clock's ability to convey information. With differing lengths of hands the number of orbits increases to 192, an information capacity of 7.585 bits. To increase the information capacity even further, more asymmetry needs to be introduced. This could be accomplished by marking the clock face with a reference mark, relative to which hand positions can be established (see Fig. 4.6). This ensures that every member of the set being acted upon by the rotation group is in a unique orbit. All 720 formal states can now be distinguished giving 9.49 bits of infor-

Fig. 4.5. Extra Asymmetry

mation capacity. Rotating the clock will not generate ambiguous time indications. I call this situation, where the number of orbits equals the number of formal states, *Case Maximum Asymmetry*.

Fig. 4.6. Case Maximum Asymmetry

A similar analysis can be performed on the error message from the deep space probe. A continuously repeating ten-bit string can be treated as a string under abstract rotation, going through the ten bits:

We have already seen that ambiguous messages can be sent; strings that are indistinguishable from others due to the rotation group acting on the set of bits. An analysis of the orbits of the group of all ten rotations acting on the set of all binary string reveals a total of 108 distinguishable states (see Appendix B). This means that if a continuously repeating 10 bit signal were to be used to indicate on-board error states, only 108 such states could be reported on, meaning that we have information capacity of only 6.75 bits. To have available all 1024 states as was intended, asymmetry needs to be introduced. The rotational symmetry can be broken by introducing a blank spacer bit at the end of each transmission of the 10-bit string. This would result in a large increase in information capacity.

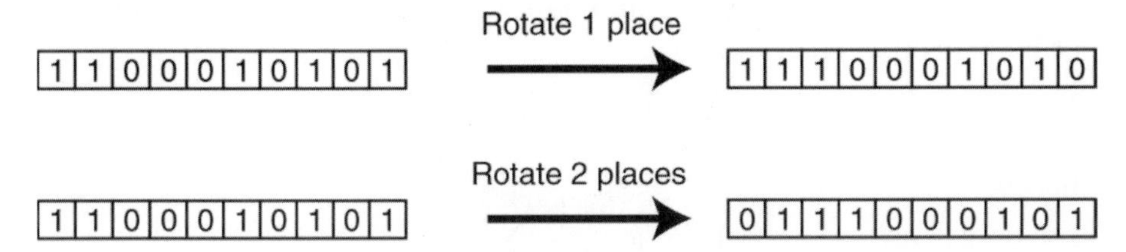

Fig. 4.7. Bit rotation

Finally, we return to the tetrahedral carbon compound to consider a slightly more complicated example. In section 3.1, we noted how enzymes use stereo-specificity to distinguish between substrate isomers. An α-carbon with any combination of 4 radicals (e.g.H–, OH–, etc.) can form a tetrahedral molecule with 28 distinguishable states (orbits). If we force each of the radicals on the molecule to be different, that is just one of each type, then there are just two orbits. These correspond to the L and D isomers of the molecule. However, most enzymes operate in a binary fashion: either a substrate is in a particular configuration or it is not. If the substrate is not optically aligned, then catalysis cannot take place. Any extra information is wasted on the enzyme.

Consider, for example, the phosphorylation of glycerol with glycerol kinase and ATP. The chemical groups attached to the α-carbon are: hydrogen, hydroxyl and two hydroxymethyl radicals and the enzyme uses a three-point binding. Despite having two apparently indistinguishable hydroxymethyl radicals, there is just one orbit: the tetrahedral molecular arrangement can only be in one state. This permits binding and the asymmetric phosphorylation of just one particular hydroxymethyl radical as follows:

$$
\begin{array}{ccc}
 & \mathrm{C^{*}H_2OH} & \\
 & | & \\
\mathrm{HO} - & \mathrm{C} & - \mathrm{H} \quad + \mathrm{ATP} \\
 & | & \\
 & \mathrm{CH_2OH} &
\end{array}
\longrightarrow
\begin{array}{ccc}
 & \mathrm{C^{*}H_2OH} & \\
 & | & \\
\mathrm{HO} - & \mathrm{C} & - \mathrm{H} \quad + \mathrm{ADP} \\
 & | & \\
 & \mathrm{CH_2{\cdot}O{\cdot}H_2PO_2} &
\end{array}
$$

This has been confirmed using carbon isotopes (indicated above using asterisks) as markers [30].

Now, at last, we are ready to finalise the formula that quantifies information transfer. The formula uses the model established in Section 3.2. In the manner of both Boltzmann and Shannon we will require that information be additive and so take the logarithm of the number of fixed orbits. Thus we have:

$$\mathrm{I} = \log(O),$$

hence,

$$\mathrm{I} = \log\left(\frac{1}{|G|}\sum_{g\in G}|S^g|\right),$$

or,

$$\mathrm{I} = \log\left(\sum_{g\in G}|S^g|\right) - \log(|G|),$$

where the base of the logarithm returns the unit of information. Choose base 2 for bits or base e for nats. This equation is the keystone relationship in my account of information. It evaluates information content of a system in terms of its asymmetries.

At this point we should summarise what has been achieved. We have seen the construction of a foundational theory of information commencing with a Leibnizian definition of distinguishability, and the relationship between information and distinguishability established. Based on this relationship, an objective, relational model has been defined which couples an informatic object with an information gathering system. The correlation between distinguishability and mathematical symmetries was established, and it has been demonstrated that information can be expressed in terms of actions of a group of symmetries, G, on a set, S. Information is quantified by the equation $\mathrm{I} = \log(\sum_{g\in G}|S^g|) - \log(|G|)$. This equation could prove to be a powerful tool with potential for application in a vast number of fields, including biochemistry and studies of causation. But for now we will focus on its role in unifying existing theories of information.

We have seen that a group theoretic approach permits us to quantify information capacity in the examples I presented earlier, but how does this stand in relation to the other quantifying theories of information? Both Shannon and Brillouin accounts of information are fundamentally probabilistic. They are effective in describing total information capacities in physical distributions. Is a group theoretic account compatible with these types of systems?

What we have developed thus far is compatible with the combinatorial account of information: the information contained in a system is related by the frequency of microstates occurring in that system. Physically, this manifests as Boltzmann and Brillouin's complexions. We can extend this to support probabilistic accounts by taking relative frequency limits of properties of collectives. That is, following von Mises' frequentist account of probability, I hold that information[7] obtained by repeated observation of attributes of an object is an objective property of the *collective* of observations, or as I shall show, a property of the object extended in time, with respect to the IGUS. To develop a conceptual basis on which to build the relationship between information Group Theory and probability distributions, I will consider a specific instance from which generalities may be drawn.

Imagine a device that consists of an opaque black box with a display on the top that is capable of showing 4 binary digits. The display is capable of being in 16 states. The actual value of the display changes at regular intervals and an observer (IGUS) makes note of these. After a large number of observations have been recorded, the records are examined. To compare the accumulated observations the IGUS requires *memory*, in this example, the recorded observations. A progressive count must be made of all the occurrences of each value until such a point that a comparison of each value may be made and the relative differences or similarities noted. In order to distinguish between the different probabilities of each value of the attributes – that is, to distinguish between the limiting relative frequencies of each of the 16 states – a store of observations must be made. In this analysis, provided that the memory is without fault, the group of values stored in memory and the observed sequence become effectively identical much in the manner of the *collective* of von Mises.

The necessity for a memory mechanism in the IGUS is critical in the evaluation of information that is extended in time (temporal information). This has been identified earlier in the consideration of Maxwell's Demon and in the work of Szilard. It is essential for the synchronous comparison of observed dynamic states. Moreover, the memory must be reliable; that is to say, it must be an accurate, permanent mapping of the accumulated observations, since it is by means of a memory image of temporal information that the IGUS has access to the observed object. Deficiencies in memory act in a manner similar to filters on the IGUS (see Fig. 3.6), restricting its capacity to apprehend in-practice distinctions in temporal maximal information from the observations.

[7] This information is formed by the probability distribution – the limit of relative frequencies of the observations.

Returning to the black box example, imagine two cases. In the first instance each of the 16 values appear with equal frequency. That is to say that the distribution of the 16 numbers is flat (see Fig. 4.8).

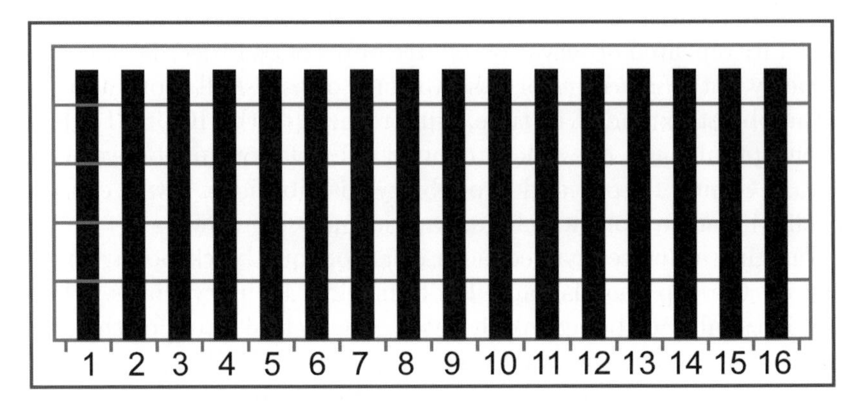

Fig. 4.8. Uniform Distribution

This indicates a highly symmetric system: no value is distinguishable from any other based on observed frequency. Inclusion of an observation symmetry transform, which we will later define, to the group acting on the system, will not increase the number of distinguishable states. No extra information has been gained. Now consider a second possibility. The observations, when charted, form a non-uniform frequency distribution as shown in Fig. 4.9.

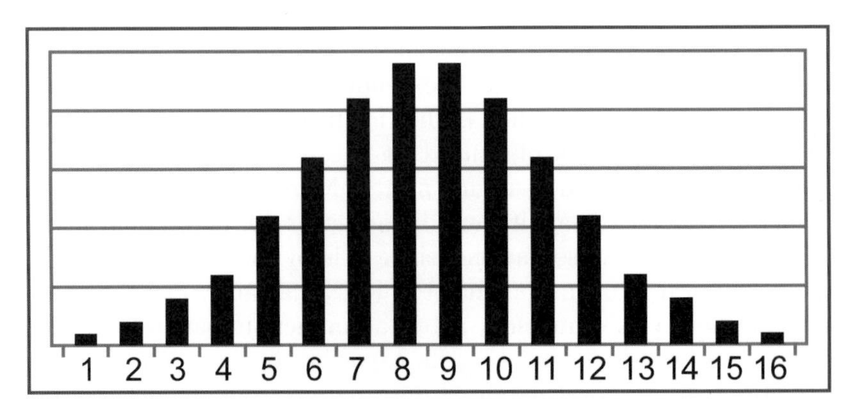

Fig. 4.9. Non-Uniform Distribution

While noticeably possessing some symmetry, it is clear that the distribution is more asymmetric than the uniform one in the previous instance.[8] Repeated observation by the IGUS of the device in this case has determined greater information capacity than the previous case. How, then, is this to be included in the group theoretic model?

Intuitively, we would like to say that the second situation is in some sense more complex; that it contains more information. If this is the case, then the increased information is embedded in dynamics since it is only through repeated observation that the asymmetry is revealed.[9] Where does the asymmetry, hence information, reside? It resides in the super-system of the box-IGUS system extended over time. Temporal asymmetry is by no means extraordinary. It is common. We find dynamic asymmetries everywhere: in the waveform of heartbeats, in transmission signals, in most dynamic processes. To see how this temporal information, based on relative frequencies, can be included in the group theoretic account of information that we have developed, let us start by determining the set of entities upon which the members of the group can act.

Let A be a set of discrete values of an attribute that can be distinguished by an IGUS. Let each $a \in A$ have associated with it a normalised value p which is the limiting relative frequency over N observations of a as $N \to \infty$ Let P be the ordered set of p indexed on a. Let G be the group of transforms which can *possibly* act on P. In previous discussion of members of a transformation group acting on material systems, the members were limited to 'physically possible transformations' that were defined as those transformations that are physically possible and able to perform a change to the system even if, in actuality, they are not implemented. The same restriction applies here with the transforms now being 'functionally' rather than physically possible. By functionally possible I mean that the transform must be capable, within the laws of mathematics, of being applied to the members of the set of P. If we limit ourselves to the scenario present in Fig. 4.9 we have A equal to the values 1 to 16. Then we can observe that since we have normalised p, all transformations of scale will be symmetries as will the application of some permutation of re-indexing P. For example, a reindexing transformation which swapped 1 and 16, 2 and 15, 3

[8] The distribution cannot, for example, remain invariant under translation.

[9] We cannot at this point say much concerning the dynamics apart from the fact that *some* dynamical process is at work. More rigorous observation may reveal increased dimensionality (complex recurring orbits) or stochasticity (statistically uncorrelated values).

and 14 etc, (a reflection about a=8.5) will also be a symmetry for all p. Other transformations will fix only some p. By analysing all possible transformations and applying the methodology developed earlier, the total distribution information relative to an IGUS can be determined.

4.3 Symmetry and Information

The relationship between Symmetry, Entropy and Information will be developed and discussed in this and following chapters. By way of preempting this process, I will summarise, in advance, my thesis regarding the relationship between these three properties. The correlation between Symmetry and Entropy is direct but inverse. The higher the number of symmetries a system is perceived as having by an IGUS, the fewer perceptible individual states the system can be in, and entropy is all about counting possible states.[10] The relationship between Information and Symmetry is also an inverse one. In the standard information theory formulation, this arises as a result of the fact that less information is required to fully describe the symmetric system. Because of their fundamental reliance on distinguishability, Information and Entropy are opposite sides of the same coin. These relationships are summarised below in Table 4.1.

Symmetry	**Entropy**	**Information**
High	Low	Low
Low	High	High

Table 4.1. Asymmetry Relationships

Note should be made here of some potentially confusing uses of the term *Entropy*. In information theory, the term entropy, H, is applied to denote the amount of self-information of a random variable or the measure of uncertainty of a random variable. As we have seen in Section 2.2.2, the entropy of a random variable X is given by

$$H(X) = -\sum_{x \in \Lambda} p(x) \log p(x),$$

where Λ is the alphabet of values x could have. The entropy of X can also be expressed as the expected value of $\log \frac{1}{p(X)}$ where X is subject to the probability distribution p(X). Thus

[10] see section 4.5.1 for elaboration.

$$H(X) = E_P \log \frac{1}{p(X)}$$

This definition is conceptually closer to Boltzmann's definition of entropy, $S = -K \log W$, where W is the thermodynamic probability. The thermodynamic probability is the number of different ways in which a thermodynamic state can be realised. Thus in an equiprobable distribution of x(that is all microstates equally likely), we have $p(X) = \frac{1}{W}$. The inverse probability relation in the second expression is manifested as the minus sign in the first. My conceptual approach to entropy will be more closely aligned with the thermodynamic second interpretation than the former (see section 4.5). For example, it has been observed that at its origin the universe was maximally symmetrical (absolutely homogenous) and of very low entropy. During the course of the evolution of this self-gravitating system, entropy increases caused the breaking of symmetries leading to greater heterogeneity and becoming more informed. It is in this sense that symmetry and entropy have an inverse relationship.

4.3.1 Information Generation

The relationship between symmetry and information is not new. Collier makes us aware of the issues at hand by considering the "paradox" associated with symmetries being both surprising and a source of redundancy:

> "On the one hand, many symmetries that we find in the world are surprising, and surprise indicates informativeness. On the other hand, the surprise value of information arises because it presents us with the unexpected or improbable, but symmetries, far from creating the unexpected, ensure that the known can be extended through invariant transformations." [24].

Though framed in terms of belief content, Collier's observations regarding the relationship between information and symmetry are nonetheless accurate. He says,

> "The information content of a belief that there is symmetry in some structure or configuration is a function of the reduction in the number of possible configurations resulting from the elimination of all the ones that are not symmetrical. The more the supposed symmetries reduce the number of possibilities, the greater the information content of the belief in the symmetries" [24].

Here we have a specific case of the general distinguishability-in-practice. Beliefs are, we suppose, based on observations of system X. If observations external to the observer-X super-system are employed in constructing beliefs about X's symmetries, then those symmetries act, in fact, on a system greater than just X. For example, beliefs may be held concerning symmetrical properties of a die where observation has been limited to just 3 sides of the object. Now, previous observations of other dice have noted rotational symmetries, so as an instance of this class, belief confers the same symmetries to the current die. However, the symmetries actually belong to the idealized class object along with the additional symmetry that is the invariance transform between each of the members of the class.[11]

Collier maintains that the relaxing or breaking of symmetries generates information in a dynamic system. In biology, biological information is produced by such symmetry breaking processes as sympatric speciation that leads to differentiation. This process is at the mercy of chance fluctuations. He also refers to proposed neurological theories of perception that speculate that the dynamics of the sense transducer and neurons are driven into a particular harmonic orbit by a sensory stimulus. If this is the case for all perception, then "perception is a form of symmetry breaking that produces perceptual information" (ibid, p.254). We have seen this (and quantified it) with the informational increase on the breaking of symmetries in our Bauhaus clock example.

Collier has illustrated instances of information in the physical world generated by symmetry breaking but he paints a much broader picture: "The original condition of the universe is statistically uniform, and hence entirely symmetrical. This statistical uniformity implies an equilibrium state (at least locally), which further implies that the early universe did not contain any information. Information, therefore, must have arisen through contingencies. The only process we know of that can produce new information from contingencies is symmetry breaking through phase separation in a system that is out of equilibrium, thereby forming branch systems. Similar branching is repeated at smaller scales as the universe differentiates and forms new branch systems" [24].[12]

[11] The invariance transform is that symmetry that makes all the dice members of the class ***Dice***.

[12] One may perhaps expect that the initial "perfect symmetry" at the origin of the universe be somewhat less than perfect and some minimal asymmetry existed since the occurrence of a phase separation is itself a break in temporal symmetry. If this were not the case, we run the risk of violating the axiomatic *ex nihilo nihil fit*.

4.3.2 Extrinsic and Intrinsic Information

Symmetry breaking makes changes to the elements g of a group G which homomorphically act on a set S such that each g of G gives us an arrangement of the points S. There is an additional, simple, though non-trivial, source of information increase available. We could change S. Consider the example of the repeating 10-bit space probe message. As it was transmitted the message was capable of sending only 6.75 bits of information. Imagine now if, without breaking the rotational symmetry, we introduced an additional bit to the message string. The message now consists of a repeating 11-bit string. A quick calculation of orbits[13] determines that the probe is now capable of sending 7.55 bits of information. Here we have increased information without breaking symmetry, but instead have simply added more crude volume. In this instance we have changed the set upon which the group transformations act rather than the group itself. With more members in the set, the number of orbits that are fixed by the action of the group increases to 188. If we can increase information in a system without breaking symmetries, what is the nature of information with regards to symmetries?

The above example may be considered to be a bit unfair. The system under consideration was physically altered, somewhat changing the rules of the game. But it does raise an important point to be considered: Information exists wholly in the physical manifestation and in the relationships of IGUS-object system. Information in the IGUS-object system illustrated in Fig. 3.6 can be thought of as existing in two modes. The intrinsic mode is primitively reliant on the physical structure of the object. It is reliant on the number of bit places, the divisions on the clock, or the atoms in the molecule. In our set theoretic representation of this system, it is reliant on the order of set S. The other mode, the extrinsic mode, is concerned with the possible configurational relations of the members of S that can be distinguished by the IGUS. This component is extrinsic in the same way that thermodynamic entropy is extrinsic. It represents the symmetry relationships of the system which are formally correspondent with the group G.

This information can be generated both intrinsically and extrinsically. Information is increased intrinsically by physically expanding the system, by directly increasing the bearing capacity. Extrinsic information is increased by symmetry breaking, as we have seen in our examples.

[13] As shown in Appendix B.

4.4 Information and Probability

Information capacity is chiefly a combinatorial problem, a counting problem in which one must determine how many states can be uniquely distinguished in an object by an IGUS. This is clear in statistical thermodynamics. The relationship between distinguishability and statistical thermodynamics is paramount. In fact, there is, as Schrödinger noted, "essentially only one problem in Statistical Thermodynamics: the distribution of a given amount of energy E over N identical systems. Or perhaps better: to determine the distribution of an assembly of N identical systems over the possible states in which this assembly can find itself, given that the energy of the assembly is a constant E" [71].

4.4.1 Maximum Entropy Principle

This approach is a corollary to the Maximum Entropy Principle(MEP) expounded by E.T. Jaynes. MEP is a formalised extension of Bernoulli's Principle of Insufficient Reason that essentially asserts:

> "(1) We recognize that a probability assignment is a means of describing a certain state of knowledge. (2) If the available evidence gives us no reason to consider proposition A_1 either more or less likely, then the only honest way we can describe that state of knowledge is to assign them equal probabilities: $p_1 = p_2$. Any other procedure would be inconsistent in the sense that, by a mere interchange of the labels (1,2) we could then generate a new problem in which our state of knowledge is the same but in which we are assigning different probabilities.(3) Extending this reasoning, one arrives at the rule:
>
> $$p(A) = \frac{M}{N} = \frac{(Number\ of\ cases\ favourable\ to\ A)}{(Total\ number\ of\ equally\ possible\ cases)}\text{"}$$
>
> [41].

Jaynes' formulation of MEP is the determination of the probability distribution in such a manner as not to make any unwarranted assumptions, leaving the maximum possible freedom, maximum uncertainty, while being subject to constraints reflecting what is known about the system.

For a discrete, single variable system x with prior information I, the mathematical problem of the MEP is to identify the $p(x|\mathrm{I})$ which will maximize the variable's entropy, H, defined by:

$$H = -\sum p(x|I) \log p(x|I)$$

subject to the following constraints:

1. $p(x|I) \geq 0$
2. $\Sigma p(x|I) = 1$
3. f($p(x|I)$) generated by I.

The problem is generally solved by the application of Lagrange multipliers.

Jaynes' approach is strongly epistemologically biased. The assignation of prior probabilities is subject to constraints of knowledge concerning the system at hand. This does not immediately transform to informatic constraint [39].

Given the epistemic nature of his approach, Jaynes' program of the Maximum Entropy Principle may be stated as follows:

1. How can we incorporate knowledge of factors which will affect a probability distribution?;
2. What form should we assume a distribution takes, if we are completely ignorant of influencing factors? How do we assign $p(x|I)$?

Once we find the prior representing complete ignorance the MEP "will lead to a definite, parameter independent method of setting up prior distributions based on testable information" [39].

Aside: As previously stated I differ from Jaynes in that I do not see entropy and information as fundamentally semantic or epistemic. However, there are sufficient useful concepts and methodology in his approach to justify its inclusion in my account.

Bayes and Jeffreys consider aspects of the second part of this problem, that of assigning $p(x|I)$. Bayes, in his revolutionary paper, *An Essay towards solving a Problem in the Doctrine of Chances*, considered the effect of knowledge concerning an event on the assignation of expected probabilities. Bayes showed that given two events, A and B, the probability that A happened given B is given by the formula:

$$P(A|B) = \frac{P(A) \cdot P(B|A)}{P(B)}.$$

Applying this theorem to the determination of P_m, Bayes described a thought experiment in which a ball is thrown randomly onto a uniformly flat and level plane, ABCD, to decide the value of P_m. He postulated, in the manner of Bernoulli's Principle of Insufficient Reason, that, when we have no prior information about P_m, we should likewise assume a uniform prior probability: equal values of dP_m. This allows us to compute the probability of the hypothesis that P_m lies between values $x_1$and x_2, given all the explicit and implicit assumptions. Laplace was influenced by Bayes and formulated the uniform distribution of priors as the *Principle of Insufficient Reason.* Indeed, Laplace held the principle to be the core of probability theory, saying,

> "The theory of chance consists in reducing all the events of the same kind to a certain number of cases equally possible, that is to say, to such as we may be equally undecided about in regard to their existence, and in determining the number of cases favourable to the event whose probability is sought" [50].

Bayes' assumption was contested by a number of people, including Jeffreys who pointed out that Bayes' uniform distribution was not invariant under a change of parameters (e.g. inversion), a crucial problem, especially if one is dealing with distributions of physical dimensional parameters. Jeffreys proposed that, for a parameter σ, a prior of $d\sigma/\sigma$ be assigned with the justification that the distribution will remain constant whether we use σ a parameter or some power function, σ^m.

> "For instance, in the law connecting the mass and volume of a substance it seems equally legitimate to express it in terms of density or the specific volume, which are reciprocals, and if the uniform rule was adopted for one it would be wrong for the other. Some methods of measuring the charge on an electron give e, others e^2; de and de^2 are not proportional. ... But while many people had noticed this difficulty about the uniform assessment, they all appear to have thought that it was an essential part of the foundations laid by Laplace that should be adopted in all cases whatever, regardless of the nature of the problem" [44].

It is not the intention here to delve too deeply into the intricacies of assignation of prior probabilities, rather to point out past attempts at incorporating notional information in probability distributions and, more importantly, to highlight similarities between the Symmetry-Group Theoretic approach presented here and the Maximum Entropy Principle of Jaynes.

Jaynes noted that the application of transformation groups could assist in the determination of priors. If a population has an original distribution of beliefs $f(\theta)$ concerning a parameter θ which is transformed to a new distribution $g(\theta)$ on presentation of some information I, then we can say that the population has learnt nothing; I was not informationally relevant to θ if $f(\theta) = g(\theta)$. We have already seen in

Section 4.1 that a transformation which leaves a parameter invariant is a symmetry and therefore contributes no additional information.

Suppose, after Jaynes, that one person, X, held that θ is the probability of a successful outcome, S, in a Bernoulli event [39]. On the presentation of additional information, I, we apply Bayes' law to determine the new expected probability θ':

$$\theta' = p(S|IX) = \frac{p(S|X) \cdot p(I|SX)}{(p(I|SX) \cdot p(S|X) + p(I|FX) \cdot p(F|X))}$$

where p$(F|X)$ = 1 − p$(S|X)$ is the prior belief in the probability of failure.

This new information has generated a continuous mapping of the parameter space $0 \leq \theta \leq 1$ onto itself by

$$\theta' = \frac{a\theta}{(1 - \theta + a\theta)} \tag{4.1}$$

where

$$a = \frac{p(I|SX)}{p(I|FX)}$$

If we apply this transformation to the distribution of beliefs in the aforementioned population and we invoke the "ignorance" condition f(θ) = g(θ), it can be shown (ibid. p.239) that the prior distribution is given by:

$$f(\theta) = \frac{const}{\theta(1 - \theta)} \tag{4.2}$$

that satisfies Jeffreys' criterion.

Now consider conducting n runs of this Bernoulli experiment with the result of r successes. The probability that we shall observe such results is given by

$$p(r|n\theta) = \binom{n}{r} \theta^r (1 - \theta)^{n-r} \tag{4.3}$$

given (4.1) and (4.2). The posterior distribution of θ is given by

$$p(d\theta|rn) = \frac{(n-1)!}{(r-1)!(n-r-1)!} \theta^{r-1} (1 - \theta)^{n-r-1} d\theta \tag{4.4}$$

In the case where r = (n-r) = 1, that is where one success and failure have been observed, equation 4 reduces to p(dθ|r,n) = dθ, the distribution Bayes took as his prior. Jaynes notes:

> "Therefore we can now interpret the Bayes-Laplace prior as describing not a state of complete ignorance, but the state of knowledge in which we have observed one success and one failure. It thus appears that the Bayes-Laplace choice will be the appropriate prior if the prior information assures us that it is physically possible for the experiment to yield either a success and failure, the distribution 2 describes a 'pre-prior' state of knowledge in which we are not even sure of that" [39].

That is to say that in the case of the Bernoulli experiment, the information concerning the nature of the experiment is built into the selection of the prior. By selecting the uniform probability $p = d\theta$ as the appropriate prior we have tacitly included information regarding the physical system: that the only possible results are success or failure. The remaining possible future states of the system, the n trials, are symmetrical in the sense that there is no additional information we can use to distinguish one of these possible states from another.

I take pause at this point to note that up to this section, the concept of information has been employed objectively without reference to beliefs or epistemic considerations. The relationship between physical information and belief structures is a complicated one that is beyond the scope of this work. It is sufficient to say that the mapping of physical information structures and their groups into knowledge structures could, I believe, be achieved by means of the distinguishability process identified in Section 3.1.

What Jaynes has shown is that by the incorporation of transformations that account for asymmetries, one can arrive a prior that is truly assumption-free. MEP then becomes the construction of a problem by incorporating all known asymmetries so that what is left is informationally minimal. By removing all transformations that result in distinguishable variance we are left with a conditionally uniform distribution of states, a set of states that is symmetric.

Jaynes' use of transformation groups and the MEP should be interpreted as a corollary to the asymmetry model presented here. Whereas I am proposing a thesis that "Symmetry denotes no information", Jaynes, from his epistemic vantage, puts forward a claim that "No information denotes symmetry". To see this more clearly, consider his treatment of Bertrand's Paradox. Proposed by Joseph Bertrand [8], the paradox illustrates the potential inconsistencies associated with failure to take into account all symmetries relevant to a particular problem and is directly related to the previously discussed Bauhaus clock ambiguities. The paradox is posed as follows: What is the probability that the length of a "randomly selected" chord to a given circle is longer than the length of a side of an equilateral triangle inscribed in the circle?

Depending on one's interpretation of "randomly selected", the answer varies. More explicitly, the choice of which parameter is to be uniformly distributed greatly affects the determined probability. If we maintain that a chord on a circle is fully described by its midpoint and that all those chords with lengths greater than a side of the triangle will have their midpoints in a circular area of radius one half the original circle, then the ratio of uniformly distributed points in the smaller circle to the total gives a probability of $1/4$ (see Fig. 4.10).

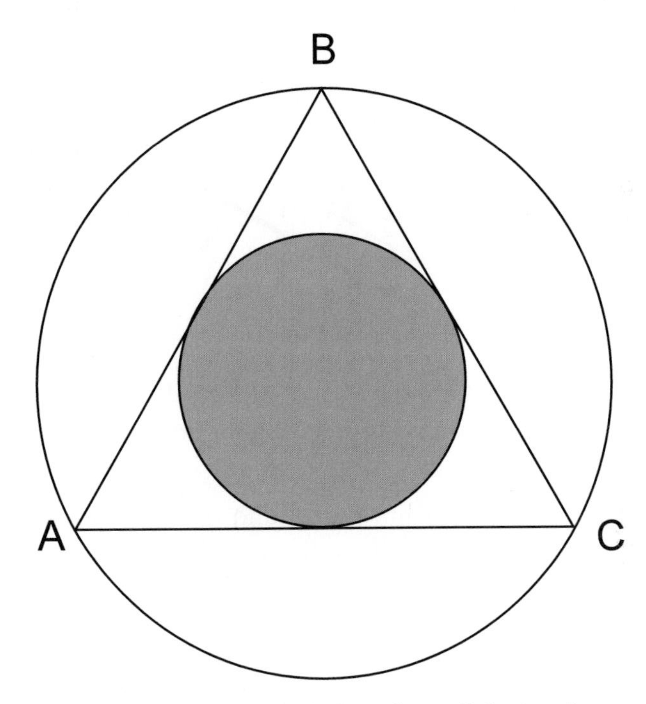

Fig. 4.10. Bertrands Paradox – Solution A

Alternatively, we could still hold that a chord on a circle is fully determined by its midpoint, but note that chords longer than the side of the triangle have their midpoints closer to the centre than half the radius. If the linear distance between the chord and the circle (along the radius) is assigned uniform distribution then the probability becomes $1/2$ (see Fig. 4.11).

Finally we can consider a chord to be determined by two points on the circle's circumference. The position of one of the points is arbitrary

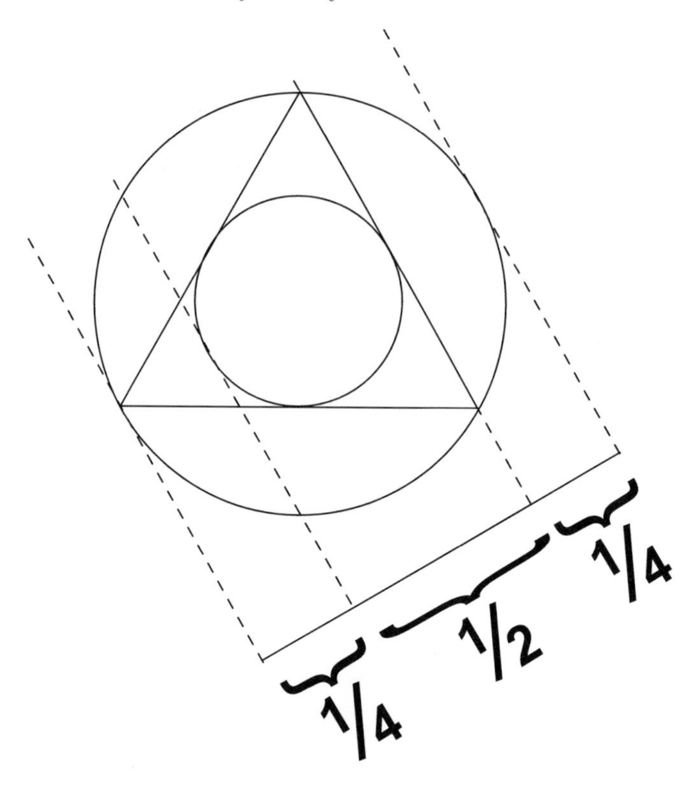

Fig. 4.11. Bertrands Paradox – Solution B

since it is only the relationship (angle and distance) between the two points that is of importance. If we fix one of the points and examine random chords emanating from that point we quickly see that $1/3$ of all chords will be longer than a side of the triangle. In this case it is the angles of intersection of the chords on the circle's circumference to which uniform distribution is assigned (see Fig. 4.12).

All three solutions seem meritorious, but which is correct? Jaynes [40] reviews 10 authors (Bertrand, Borel, Poincaré, Uspensky, Nortup, Gnedenko, Kendell and Moran, von Mises and Mosteller) examining the paradox. Of these, only Borel indicates a preference (that the majority of natural processes point to solution B, though he provides no proof). Von Mises declares that the problem does not belong in probability theory. The remainder, noting that the allocation of uniform distribution is at the heart of the problem, maintains that there is no definite solution because the problem is ill posed. The tendency, then, would be to nominate that the only reasonable method for the assigna-

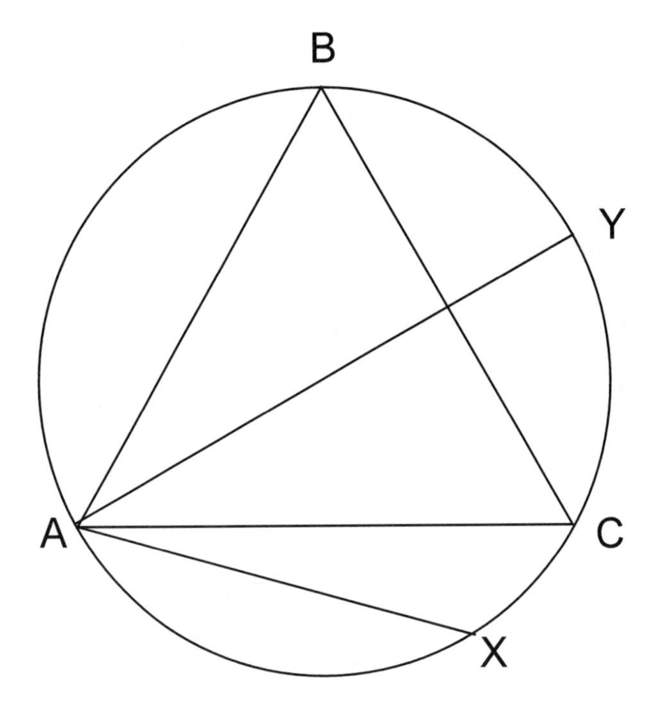

Fig. 4.12. Bertrands Paradox – Solution C

tion of probabilities is to adopt the frequentist approach and conduct random experiments.

Jaynes, however, maintains that often it is possible to allocate *a priori* probabilities given informationally restrictive problems "allowing many different solutions with nothing to choose among them" [40], just as it has been done in physics.

> "For example, given the average particle density and total energy of a gas, predict its viscosity. The answer evidently depends on the exact spatial and velocity distributions of the molecules (in fact, it depends critically on position-velocity correlations), and nothing in the given data seems to tell us which distributions to assume. Yet physicists *have* made definite choices, guided by the principle of indifference, and they *have* led us to correct and non-trivial predictions of viscosity and many other physical phenomena" [40].

The procedure for doing this is to identify the probability function or functions that are invariant under possible transformations relevant to the designated problem. The candidate function or functions are constrained by the information associated with the problem. So, with his characteristic epistemic tilt, Jaynes' restatement of the problem becomes: "Which probability distribution describes our *state of knowledge* (his italics) when the only information available is that given in the above statement of the problem?"(ibid).

From the specification of Bertrand's Paradox, Jaynes identifies three possible transformations concerning which no information is given. These are: rotational (no angular position is specified[14]), scale (no circle size specified) and translation (no location specified). If "the problem is to have any definite solution, it must be indifferent to these circumstances" [40].

Jaynes examines these transformations individually and determines the constraints that they impose on candidate solutions. Rotational symmetry is irrelevant to the distribution of chords, as an observer's vantage makes no difference to the distribution. All three candidate solutions described above are invariant under rotation.

Scale invariance poses a tighter constraint on the problem. Specifying that the probability density remain the same for circles of different sizes means that if the angles of intersection were distributed uniformly on one circle, they would not be uniformly distributed for a smaller, concentric circle. Thus candidate solution C is eliminated. The other two proposed distributions satisfy the scale constraint.

The requirement of translational invariance further limits the range of possible solutions. As the uniform distribution of chord midpoints over the interior of the circle is not invariant under translation, solution A is eliminated. By considering the invariance constraints, Jaynes analytically determines the probability distribution that should be:

$$f(r, \theta) = \frac{1}{2\pi R r} 0 \leq r \leq R,\ 0 \leq \theta \leq 2\pi$$

where r, θ are the polar coordinates of the chord and R is the radius of the circle.

Solution B remains the only solution that satisfies this function.[15]

At the heart of the issue is the relationship between information presented (or not presented) and symmetries. If in the original problem no information was presented concerning a particular possible transformation, then we must assume that this transformation is symmetric. The failure here to fully specify the problem allows ambiguities and so invariance must be assumed. In this manner it is possible to find

[14] As in the Bauhaus clock example.

[15] Jaynes claims to have experimentally verified this result using broom straws and a 5-inch circle with 128 trials (ibid.).

a unique, definite solution. Other problems, however, may be more ambiguous and underdetermined making it impossible to decide on a parameter to assign a uniform distribution. If the problem were fully defined, all relevant information specified, then we would have a system with Case Maximum Asymmetry as specified in Section 4.2.2

A problem may appear to arise at this point. If we assume that there is information that has not been provided, that is there are properties which are distinguishable in principle but not in practice (within the terms of this experiment), then this potential information will not manifest but rather will be hidden as symmetries. So when constructing the probability function exactly which of all the possible transformations should be taken into consideration? There are presumably infinite possible transformations unspecified. In this example Jaynes considers three kernels (rotation, scale and translation) in his transformation group. Not specified are such possible transformations as time of day, colour of circle, etc. By inspection we realise that the final probability distribution will remain invariant under transformations such as these. How are we to deal with subtle but germane transformations? Jaynes' acknowledges this is an issue but is unclear on an answer. He appeals to rationality when specifying problems: "In the first place, we recognize that every circumstance which our common sense tells us may exert some influence on the result of an experiment ought to be given explicitly in the statement of a problem" [40].

If we undertake a calculation to determine a probability function as described above, we are not entitled, we are told, to assert that the distribution will be observed in practice. If an experiment is performed with the intention to verify the prediction and the expected distribution is not forthcoming, then we can conclude that at least one of our invariance assumptions is wrong, that an additional asymmetry exists, which represents, on my account, additional physical information not included in the problem description. On initial inspection this may appear to be challenge to Jaynes' tendency to Idealism, however Jaynes may reply that this result just represents our imperfect knowledge regarding the state of things. However, my approach is unaffected by this, since, at the very least, the discerning of objective relationships precedes the construction and manipulation of knowledge structures and it is at this lower level where information exists.

The selection of which symmetries to include is a pragmatic consideration. The inclusion of all possible transformations consistent with the specified scenario will include those germane transformations that will affect the possible distribution set and those irrelevant transformations that will be symmetries and leave the distribution solution unchanged; however this may well prove unfeasible from a practical standpoint.

The discussion of Jaynes' Maximum Entropy Principle over the past few pages has focussed on its application to epistemic problems, well-posed or otherwise, rather than directly at physical information-bearing system. However, the analysis of Jaynes has been useful and relevant as a corollary to the information-as-asymmetry model I am proposing here in that it brings into relief the absence of information that symmetries entail. It is perhaps also useful to examine a physical system correspondent with the problem set forth in Bertrand's Paradox and examine the physical information in accordance with the schema we have developed thus far.

Consider, as Jaynes proposes, a rain of thin straws over an arbitrarily large area A. Within this area A there is a circle of radius R such that $\pi R^2 \ll A$. And consider a length $\sqrt{3}R$ (this is the length that a side of an inscribed equilateral triangle would be, were it drawn).

In representing the specified problem and the information provided (or not provided), we can consider an IGUS apprehending information from the straws/circle system. The IGUS has "filters" on the maximal information from the straws/circle system such that the IGUS cannot distinguish in practice the orientation of the circle, the size of the circle and the position of the circle.

Firstly, we note that there is nothing to distinguish the rotation of the circle, since, as with the Bauhaus clock the angular position of the IGUS relative to the circle is immaterial: the distribution must be the same regardless of the coordinates of the IGUS.

If we assume the rain of straws to be random in such a manner that the angle of incidence of the straws on the circle was uniformly distributed, then small changes in R (around the same centre) would be distinguishable by the IGUS since the proportion of straw section greater than $\sqrt{3}$R would vary. Here, depending on how we construct this example, two outcomes are possible. If we maintain that, by definition, the IGUS cannot in practice distinguish scale information, then our assumption must be wrong. If, however, we stand firm on our definition of randomness (perhaps due to the manner in which the rain was generated), then we must allow the ability to distinguish variation in scale of the circle.

Wherein lies the information? Here, the information arises from the asymmetric relationship between the straws and the varying circle sizes. If the angles are distributed uniformly, the "random" rain of straws embodies information about the circle's size in so far that the arrangement of straws is associated with one specific circle size. If the intersections of chords on the circumference of a particular circle, C_1, were arranged in such a manner as to generate a uniform distribution of angles, the chords would not be distributed in angular uniformity with respect to a concentric smaller circle, C_2, drawn inside C_1. It is perhaps of interest to note, however, that the correct combination of both a scale transformation and a translation will produce the same distribution for two different circles. Consider Fig. 4.13 below. Increasing a circle's radius *and* translating it in such a manner that the circles touch at point A will ensure that the uniform distribution of straws based on angle of incidence will be the same for both circles.

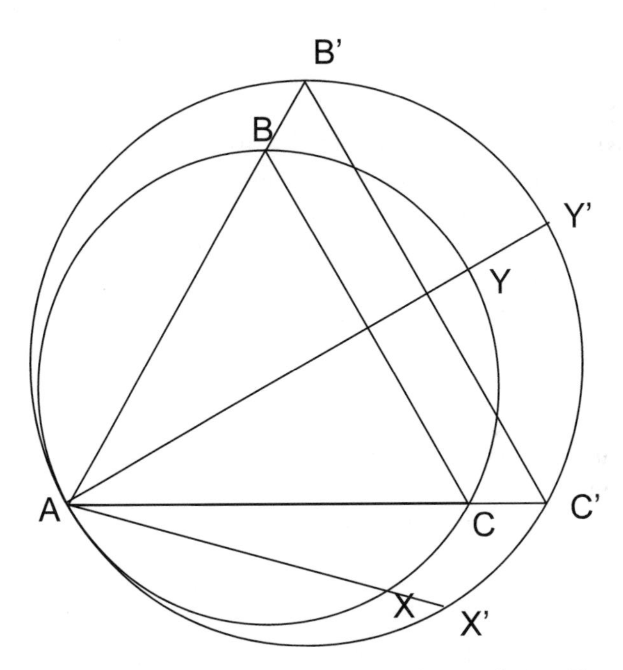

Fig. 4.13. Bertrand's Paradox with Scale and Translation Transformations

This case of embedded information also applies for transformations of coordinates, or translations. The distribution of angles of incidence of chords on the circle circumference is strongly linked to the position of the circle. Further, the uniform distribution of the chord midpoint over the centre of the circle, is also linked to the circle position, so that a change in location will, *ceteris paribus*, be associated with a single distribution of the chord midpoint over the centre of the circle.

The notion that MEP can be used as an approach to define information has been put forward by several writers [45]. However, these definitions of information are typically subjectivist and epistemic in nature with an emphasis on informatic meaning and semantics. This is to be expected given Jaynes' standpoint. There is much that is problematic in Jaynes' account of Statistical Mechanics. Jaynes maintains that Statistical Mechanics be based on a probability that is subjective. That is, a description "of a certain state of knowledge" [41], and on a calculus of inductive reasoning as described in his unpublished book *Probability Theory: The Logic of Science* [43]. However, it is not my task here to develop a thorough critique of Jaynes' work nor is it to tease out the numerous conceptual and technical problems with Jaynes' account of Statistical Mechanics. I attempt rather to find that which can be re-purposed for an objective account of information and entropy and Jaynes' MEP approach *does* capture the asymmetry 'essence' of information.

Consideration of Bertrand's Paradox as an IGUS-object system has shown that physical systems can be analysed by the application of transforms and the resultant asymmetries describe the embedded information. Jaynes' Principle of Maximum Entropy, though from an epistemic view point, still holds in an objectivist realm and the isolation of all asymmetries guarantees the remaining symmetries have no informatic content.

4.5 Information and Statistical Mechanics

4.5.1 Distinguishability and Entropy

Distinguishability lies at the heart of statistical-mechanical theories concerning entropy. The determination of the number of states a system may possibly manifest subject to the physical constraint and the combinometric evaluation of these states to quantify entropy is the essence of statistical entropy theory; it is all about complexions. By way of illustrating this concept consider the determination of the entropy of an ideal gas with N atoms.[16] We have seen already (Section 2.2.1) that in his response to Loschmidt's *Umkehreinwand*, Boltzmann proposed the following definition of entropy:

$$S = -K \log W$$

[16] Ideal gases are ones that follow Boyle's Law (PV=nRT) exactly.

where K is a constant and W (the thermodynamic probability) is calculated as follows:

$$W = \frac{N!}{\prod_i N_i!}$$

By considering the translational properties of the gas atoms under a Boltzmann-Maxwell distribution and the effects of quantised translational levels (that is, quantum degeneracy), it is possible calculate an equation of state for the gas and then determine the theoretic entropy.[17] This entropy for an ideal gas is calculated to be:

$$S = K \log N! + NK \log \frac{(2\pi m)^{\frac{3}{2}} (kT)^{\frac{5}{2}}}{ph^2} + \frac{5}{2} NK$$

where: K is Boltzmann's constant;
N is the number atoms in the system;
m is the atomic mass;
T is the temperature;
p is the pressure;
and h is Planck's constant.

Having determined this entropy we are now at liberty to test the theory against a real monatomic gas, such as neon. We can experimentally infer the absolute entropy of a gas by taking S = 0 at T = 0 Kelvin (The Third Law of Thermodynamics. See section 4.6.2), measuring physical properties required by the following expansion of Clausius' original formulation:

$$S = \int_0^{T_{\text{fus}}} \frac{C_S}{T} dT + \frac{\Delta H_{\text{fus}}}{T_{\text{fus}}} + \int_{T_{\text{fus}}}^{T_{\text{vap}}} \frac{C_L}{T} dT + \frac{\Delta H_{\text{vap}}}{T_{\text{vap}}} + \int_{T_{\text{vap}}}^{T} \frac{C_G}{T} dT$$

where: C_S is the solid phase heat capacity;
C_L is the liquid phase heat capacity;
C_G is the gas phase heat capacity;
ΔH_{fus} is the heat of fusion at temperature T_{fus};
ΔH_{vap} is the heat of vaporization at temperature T_{vap}. Using this method of verification, we find that a discrepancy exists between the predicted and measured entropies with the former being much larger than the latter. The theoretic value is, in fact, too great by the amount $K \log N!$. There appears to be a serious problem with using the Boltzmann-Maxwell distribution, as it appears not to provide an accurate entropic account, even for monatomic, idealized gases. What is missing?

---

[17] see, for example, [84].

The problem lies in the implicit assumption that all particles are permanently distinguishable from each other. This is not the case. No account of atomic degeneracy[18] is taken in this approach as consideration is given only to the translational degeneracy (momentum distributions). The number of different ways that we can take N distinguishable particles is $N!$. If each atom were completely distinguishable from all others, then by Boltzmann's H-theorem, this would account for the additional $K \log N!$ term. Apart from disagreement with empirical results, the Boltzmann-Maxwell approach has some unpleasant conceptual implications. If all atoms were distinguishable, the addition of 1 mole of atoms of a gas A at temperature T and pressure P to another mole of A at the same temperature and pressure would require an increase in entropy upon mixing. This would mean that entropy is not an extensive property that is determined solely by the state of the system.[19]

The removal of the $K \log N!$ term rectifies the situation and expression for the entropy of an ideal gas becomes:

$$S = NK \log \frac{(2\pi m)^{\frac{3}{2}}(kT)^{\frac{5}{2}}}{ph^2} + \frac{5}{2}NK$$

This equation is known as the Sackur-Tetrode equation after Otto Sackur and Hans Tetrode who contributed to the formulation in 1912 and demonstrated the need for quantisation in classical gas laws.

Application of the Boltzmann-Maxwell distribution is generally inappropriate in systems where degeneracy is present. Attempts to apply the distribution to quantum mechanical particle systems where spin is important have failed. In boson systems[20] we use an ensemble where the only possible states are those invariant under permutation of the identity of the particles. This is called the Bose-Einstein distribution and results in a reduced number of permutations with respect to the Boltzmann-Maxwell distribution. Similarly, in the case of systems of particles with antisymmetric eigensolutions (fermions) that are not permanently distinguished from each other the Fermi-Dirac distribution is employed. The distribution is as the Bose-Einstein distribution but a sign change is employed for odd permutations.

At this point we can now allude back to Jaynes' maximum entropy approach discussed in Section 4.4.1. With the application of the

[18] Degeneracy refers to the number of distinct probability distributions for a system which all have the same energy level.

[19] This is reviewed in detail in our consideration of Gibbs' Paradox (Section 4.6.3).

[20] Systems composed of entities with whole number spin and with eigensolutions symmetric in character.

Boltzmann-Maxwell distribution to the calculation of the entropy of an ideal gas, we made no assumptions concerning symmetries present in the system about which we had no knowledge.

I have previously noted that high entropy denotes high information. It is important to be very precise about what is meant by this as it may seem somewhat paradoxical to insist that the more random a system is, the more information it can contain. Throughout this work I have been focussing on information capacity, the amount of information that can be distinguished in practice by an IGUS. In the consideration of symmetry and information capacity of physical systems we must be very specific about that of which we speak.

Consider the following process of the phase transitions under heating of a solid crystal of material X to liquid then gaseous states observed by an IGUS Y. The process goes from a low entropic state to a high entropic state. The information that may be passed from the X to Y according to the schema described in Fig. 3.6 depends fundamentally on the ability of Y to distinguish in practice different states of the system composed of X. That is, the greater the ability of Y to distinguish different states of X, the larger the subset of the total states distinguishable in principle and hence the more information imparted. Now, consider that Y, much in the manner of an unaided human, was unable to distinguish microstates (that is at the atomic level) of X. The observable properties of the crystalline X would appear uniformly distributed at the macrolevel. After melting to a liquid, there may still appear to be little to distinguish the still apparently uniform property distribution, though the actual melting phase-transition point will provide an asymmetry. On subsequent heating, macrodistinguishable features such as Bernard cells appear representing an increase in asymmetries. When X has become a gas, the macrodistributions are again predominantly uniform.

Now consider the case where Y is capable of perceiving properties at the atomic energy level. In this case when X is in its crystalline form, Y will not be able to distinguish one atom from the others because of their being bound in a regular repeating structure. Barring major lattice flaws, translational symmetry is high. In the liquid form, the crystalline symmetry is broken and there are many sets of translational energy states that Y can distinguish from each other. In the gaseous state the energy sets are more numerous so the number of ways the system of X atoms can be in the states is greater. Therefore the information capacity is greater. If Y is capable of distinguishing microproperties, the information that the system of X can convey is proportional to the number of microstates. So is entropy contingent on the facility of an IGUS to distinguish states?

The entropy to which we refer in Statistical Mechanics is Case Maximum Asymmetry. That is, it is the number of states that can be distinguished in principle or, at least the number of the states that the microentities themselves can distinguish.[21] In this sense, it is entropy with respect to the object itself. If we assign proportionality to the relationship between information and entropy, then we will need to specify the conditions under which the entropy is defined. Entropy then would become entropy-with-respect-to something. This, however, seems unsatisfactory. Although entropy is an extrinsic property, we should still consider it to be a property of the system independent of the nature of an IGUS. We can get this quality if we assign general entropy to be, as stated above for Statistical Mechanical entropy, case maximal asymmetry where the total number of complexions is that which can be distinguished by the components themselves.

This will be discussed in Section 4.6.1 where we will see that entropies depend largely on the context in which they are to be used.

4.5.2 Demonic Information

It is clear in the previous example that the Y that could distinguish microstates was akin to a Maxwellian Demon. We are currently in a position to re-visit the Brillouin-Szilard interpretation of the relationship between information and entropy. In the construction of the Demon thought experiment, the Demon is an IGUS which is capable of distinguishing velocities of molecules, or, more precisely, capable of making a binary distinction of whether a particular molecule is greater or less than a particular velocity, say V. We have already seen in Section 4.2.2 that a memory mechanism internal to the IGUS is crucial to store the distinctions. Here the Demon needs at least a binary-state memory to register a single molecule's velocity (greater or less than V). On allowing or denying the passage of the molecule the Demon expends $kT \ln 2$ of energy.[22] Before the next molecule can be assessed for sorting, the internal binary memory of the Demon must be reset. According to Landauer [49] and later Bennett [5], the transformation of any logical computational memory state to an erased one is a many-to-one mapping which has no unique inverse; it is logically irreversibility. In symmetry

[21] One might argue that there may be other features that are distinguishable such as quantum differences. If this is case then the problem domain has not been correctly posed.

[22] About $3\text{x}10^{-21}$ Joules.

terms the internal state of the Demon prior to measurement, that of ignorance, is a highly symmetric one. Depending on how one represents this internal state, this symmetry may manifest as uniform probability distribution across a number of possible states, or as a binary string of a number of zeros. In the measurement of a molecule in this binary state system, the number of internal states of the Demon doubles from the symmetric 1 to the less symmetric 2. Symmetry breaking has occurred and information has been gathered.

But what of degrees of perceptiveness? In our model of the Demon/IGUS we have allowed (see Fig. 2.3) for the possibility of various degrees of passive filtering, filtering which governed the degree to which the IGUS was capable of distinguishing in practice the totality of what was distinguishable in principle according to the physics of the situation. What would be the effect of a Demon that was capable of distinguishing molecular energies at a granularity finer than that of just greater than V or less than V? Surely the information transferred from the gas system to the Demon would be greater than that with a Demon with the mere ability of distinguishing binary states.

Consider the binary state Demon. The total entropy increase for the Demon on sorting N molecules would be $N.k.\ln 2$. If the Demon were capable of distinguishing M different states of the energy of a molecule then the total entropy would be $N.k.\ln M$. If the Demon were capable of distinguishing *every* molecular complexion , then the total entropy would be $N.k.\ln \Omega$ where Ω is the total number of distinguishable states. As the degree of perception becomes more acute and the number of distinguishable states increases, it is clear that the entropy increase of the Demon also grows directly due to measurement. This is due to the amount of internal asymmetry required to store the state of the examined molecule.

But do we get any extra work for this extra entropy? As will we see from considerations of Gibbs Paradox in Section 4.6.3, increased acuity does yield increased work. But for now I wish to note that if a Demon were capable of distinguishing case maximal asymmetry, that is of distinguishing Ω states (taking in to account the symmetry of individual molecules as noted above) then the entropy of resetting the memory of the Demon, that of collapsing internal asymmetries to just one, would equal the total entropy of the gas relative to the absolute zero as defined by the Sackur-Tetrode equation.

4.6 Information and Physical Thermodynamics

4.6.1 Symmetry and Physical Entropy

Thermodynamic entropy is a slippery concept, one whose meaning depends very much on what you are talking about. In the field of chemistry, there exist many forms of entropy: entropy of mixing, entropy of reaction, entropy of melting and so forth. Here we will examine some examples of these entropies and consider the role symmetry plays in their nature.

While atoms have one internal degree of freedom – electronic – molecules have many degrees of freedom – electronic, rotational and vibrational – which each contribute to total energy, entropy and to other macroscopic thermodynamic properties. If a physical system has a high information carrying capacity, that is the system has such a symmetry structure as to have many internal states capable of being distinguished, then we would expect physical entropies higher than energetically comparable systems with more symmetry. Organic chemistry affords many examples of just such physical systems in the form of isomers.

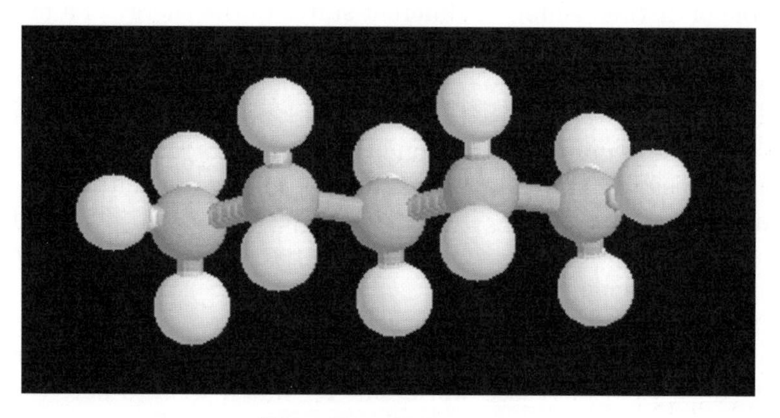

Fig. 4.14. N-Pentane

Physical properties of chemicals are obviously determined in part by the geometry of their constituent molecules. The cohesive strength of molecules comes from van der Waals forces. This internal strength governs such properties as density, boiling and melting points etc. Long chain molecules pack together more solidly than their branched or ball-like isomers. This leads to chain-like molecules such as n-pentane having higher bulk densities (0.6262 g/ml) than a branched isomer like isopentane (2-methylbutane) (0.6201 g/ml), which in turn is denser than the tetrahedral neopentane (2-2-dimethylpropane) (0.6135 g/ml) which packs loosely due to the interstices.

Fig. 4.15. Isopentane

These isomers also have boiling points which decrease in value: n-pentane 309.22 K, isopentane 301.00 K and neopentane 282.65 K. The melting points, however, behave in a different manner. The melting point of n-pentane is 143.43K, and isopentane melts at 113.25K however neopentane breaks the trend and melts at relatively warm 256.6K, 114 degrees warmer than its linear isomer. What can explain this anomaly?

The answer is symmetry. As energy is put into a solid-state system approaching its melting point, the molecules manifest increased kinetic energy. We must consider what sorts of motion the molecules are capable of as the packed solid state undergoes melting. The various freedoms of kinetic energy that are possibly adopted are translational motion, rotational motion and conformational motion.[23] On melting, neopentane molecules gain only one additional degree of freedom of motion, that of translation. This is because tetrahedral/spherical molecules such as neopentane are capable of rotational motion even in the solid state and are highly rigid so that they are incapable of much conformational motion regardless of the state. The different symmetries are distinguishable by the molecules themselves, manifesting as differences in possible kinetic phase spaces.

[23] That is, internal motion of the atoms in the molecule.

Fig. 4.16. Neopentane

Chemists and Physicists classify molecules according to their symmetry and assign them a *symmetry number*, σ. A molecule with high symmetry number such as methane or benzene (σ12) has a reduced entropy associated with melting. This decrease is of the order of $R\ln(\sigma)$ relative to the asymmetrical analogue molecule. Since the enthalpy of melting is related to the entropy of melting and the temperature of melting by the equation $T_{\mathrm{melt}} = \Delta H_{\mathrm{melt}}/\Delta S_{\mathrm{melt}}$, a change in symmetry that reduces ΔS_{melt} without affecting ΔH_{melt} will cause T_{melt} to increase. That is to say that a symmetric molecule will have a higher melting point than an asymmetric isomer, because it has a lower entropy in the melt. Examining the ΔS_{melt} values for the pentane isomers, we see this is borne out. N-pentane has an entropy change of 58.6 joule/K on melting, 45.2 joule/K for isopentane and 12.6 joule/K for neopentane. Here we see physical manifestation of the inverse relationship between symmetry and entropy.

4.6.2 Symmetry and the Third Law

We have seen that entropy is a relative concept with changes in entropy being of primary interest. In Section 2.2.1 we briefly mentioned Nernst's theorem or the third law of thermodynamics, which sets the absolute value of entropy. The theorem states that *as the temperature diminishes indefinitely, the entropy of a chemically homogenous body of finite density approaches indefinitely near to the value zero.* That is, as temperature approaches absolute zero for a chemical homogenous body, entropy approaches zero. For our analysis of symmetry, information and physical entropy, the important term here is "chemically homogenous body". Planck's formulation of the third law states, *the entropy of a perfect crystal at absolute zero is equal to zero.* To see why this is important let us turn back the problem of M. Van Der Hooft's diamonds. Let us assume that the two diamonds are structurally perfect and ask the following question: If we took both the Hofmann and the Bloit diamonds and were somehow able to cool them to absolute zero would we be able to say that Bloit's pure C_{12} diamond would have the same zero entropy as Hofmann's mixed isotope diamond? The answer depends on whether the asymmetry induced by isotopic distinguishability is at issue for the system itself. Fredrick Wall notes,

> "A solid solution that is nearly perfect but still has a positive entropy, even at absolute zero, is a crystalline mixture of isotopes. Under most circumstances, the chemical reactivities of isotopes can be regarded as equivalent, and hence two or more isotopes, in either elemental or compound form, will tend to be randomly distributed among the sites available. Taking cognizance of the existence of the isotopes will then require one to assign a positive entropy to the system. Practically speaking, however, one can disregard the entropy of mixing of the isotopes provided the entropies so calculated are used only in connection with the reactions in which no separation of isotopes occurs. Since separation of isotopes is difficult to attain, especially through chemical reactions, one can in practice forget about the existence of isotopes without introducing appreciable error in thermodynamic calculations. For processes that do give rise to separations, however, full cognizance must be taken of the entropy changes attending to isotopic mixing" [84].

Structural asymmetry similarly can induce non-zero entropies at absolute zero. As Wall notes, "In super-cooled liquids or glasses at absolute zero one can expect entropies greater than zero. Since a liquid does not possess the order that is characteristic of a crystal, it will have a positive entropy, which can be regarded as 'frozen-in' when the liquid is subjected to super-cooling" [84].

4.6.3 Information and The Gibbs Paradox

The model that we have developed in this thesis with regards to distinguishability and information can aid us in understanding a classic 'paradox' in thermodynamics, namely the Gibbs' Paradox (after J Willard Gibbs). As a corollary to Maxwell's Demon, the experiment involves the measuring the entropy produced on the mixing of two gases. Consider a gas with particles labelled A and B that is divided into equal volumes V and molecular numbers N separated by a partition. The partition is removed and the molecules are allowed to mix. If we calculate the change in entropy for the A particles we find:

$$\Delta S_A = kN \ln(2V) - kN \ln(V) = kN \ln 2$$

A similar calculation for the B particles determines $\Delta S_B = kN \ln 2$ also. The total $\Delta S = 2kN \ln 2$. However, if we consider entropy to be an extensive variable we note the entropy of 2N molecules in 2V volume is twice that of N molecules in V volume. So the net entropy change$\Delta S = 0$. Given our considerations of symmetry and distinguishability we can readily say that this apparent 'paradox' is due to the mutual distinguishability or otherwise of the molecules. If the molecules can be distinguished then the first ΔS value holds. If all the molecules are indistinguishable then there is no change in entropy. To clarify this, imagine what is required to reverse the mixing process. If the molecules are identical then no energy is required the return the system to the initial state; the partition simply needs to be reinserted. If, on the other hand, the molecules were distinguishable, then a Maxwellian-like Demon would be required to separate (or sort) the molecules.

Jaynes considered the following thought experiment based on Gibbs Paradox to illustrate the fact that more information and high degrees of discrimination can generate more thermodynamic work. Jaynes imagines two types of Argon gas A1 and A2. With our present capacities we are not able to distinguish between A1 and A2 in practice. Experiments where we mix equal molar quantities of A1 and A2[24] lead us to the conclusion that there is no change in entropy on mixing.

But A1 and A2 are distinguishable in principle, and we can imagine that in the not too distant future we arrive at a technological point where we are able to detect a difference between the two types of Argon, namely that A2 is soluble in Whifnium (a rare superkalic element we

[24] Note that given we can't distinguish between A1 and A2 it is only by the greatest of improbabilities that we would have in our possession pure quantities of A1 and A2.

are told) whereas A1 is not. The A1-A2 mixing experiment is repeated with $n_1 = fn$ moles of A1 in the volume V_1 and $n_2 = (1 - f)n$ moles of A2 in the volume V_2. On mixing the entropy increase ΔS is given by:

$$\Delta S = -nR(1 - f) \log(1 - f)$$

Jaynes notes:

> "But if this increase is more than just a figment of our imaginations, it ought to have observable consequences, such as a change in the useful work that we can extract from the process ... The amount of useful work that we can extract from any system depends – obviously and necessarily – on how much 'subjective' information we have about its microstate, because that tells us which interactions will extract energy and which will not; this is not a paradox, but a platitude" [42].

Jaynes proceeds to describe an experiment to realise this work. Starting with the original construct of Gibbs Paradox, let gas particles on one side of the partition be A1 and the other be A2. When the partition is removed, the molecules of A2 will diffuse through the piston until the partial pressure of A2 is the same on both sides. The piston is allowed to move under isothermal expansion in the direction of increasing V_1. The work done by the expansion of A1 is

$$W_1 = \int_{V_1}^{V} P_1 dV = n_1 RT \log(\frac{V}{V_1})$$

or

$$W_1 = T\Delta S_1.$$

Imagine now that the corresponding superkalic element Whafnium is discovered which is permeable to A1 and not A2. We can repeat the above experiment with a Whafnium piston and extract $W_2 = T\Delta S_2$. Combining these two processes into a dual-piston apparatus we can extract a total $W = T\Delta S$ which corresponds to the same drop in free energy that is predicted by thermodynamics. It is clear that the capacity to distinguish between microstates, to have a new degree of freedom, has enabled us to extract work based on changes in entropy. Were it to be the case that there exists, at a finer level, another in principle distinction, an additional degree of freedom, and if this become a distinction in practice, we could expect that even further work could be extracted. Imagine, for example, there are two types are Argon A2, say $A2_a$ and $A2_b$ which are distinguished by $A2_b$ being soluble in yet another superkalic element, Whoofnium. A similar apparatus might be constructed and further work extracted. And so on ad infinitum. Here we can close our discussions of the previous section on Maxwellian Demons and note that we have proof that increased acuity does have rewards.

The increasing degrees of freedom that accompany distinguishability in practice advancing closer to distinguishability in principle does not of course affect the veracity of the laws of thermodynamics. The inclusion or otherwise of groups of symmetries in a thermodynamic construct merely serves to define the scope of the problem. Jaynes notes:

> "[E]ven after the discovery of the superkalic elements, we still have the option not to use them and stick with the old macrovariables $\{X1 \ldots Xn\}$ of the 20th Century. Then we may still ascribe zero entropy of mixing to the interdiffusion of A1 and A2, and we shall predict correctly, just as it was done in the 20th Century, all the thermodynamic measurements that we can make on Argon without using the new technology. Both before and after discovery of the superkalic elements, the rules of thermodynamics are valid and correctly describe the measurements that it is possible to make by manipulating the macrovariables *within the set that we have chosen to use*" [42].

That is to repeat that thermodynamic entropy is a slippery concept; one of those whose meaning depends very much on what you are talking about.

Through the consideration of The Gibbs Paradox we have noted the objective relationship between information and work. The greater the ability to distinguish between states, that is the more information an IGUS has, the greater the work that can be extracted. In the next section we consider the relationship between Algorithmic Information Theory and symmetry.

4.7 Quantum Information

With almost a century of study into quantum physics and with the recent advancements in the field of quantum computation, it would be remiss not to examine, if only briefly, the nature of quantum information and how it sits with a group theoretic account of information. This is even more true given the special characteristics of quantum phenomena.

4.7.1 Quantum Information and Distinguishability

The ability to distinguish systems which differ in quantum mechanical properties represents a capacity that is entropically significant. We noted in Section 4.5.1 that the application of the Boltzmann-Maxwell distribution is generally inappropriate in systems where degeneracy is present and that attempts to apply the distribution to quantum mechanical particle systems where spin is important have failed. Distinguishabililty, we have seen, is inextricably tied to measurement . However, it is at the very heart of quantum systems that the accuracy of direct measurement of atomic and sub-atomic phenomena is bound by Heisenberg's uncertainty principle which states that if you determine the momentum of a particle with an uncertain of Δp you cannot determine the position with accuracy greater than $\Delta x = h/\Delta p$. What we can measure though is frequency distributions from experiments on quantum systems and determine probabilities from them.

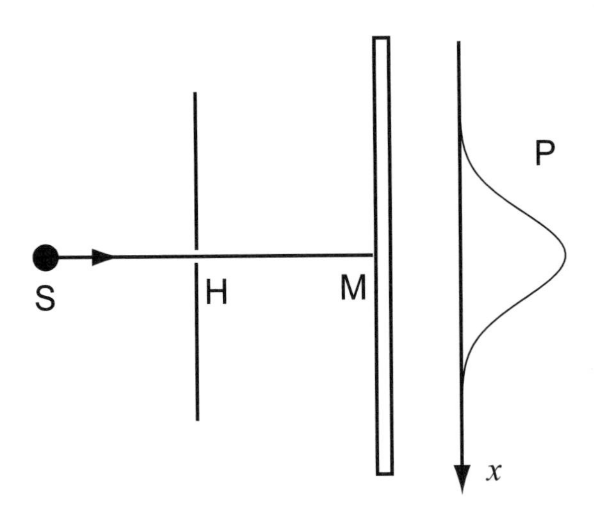

Fig. 4.17. One Hole Experiment

To illustrate the special considerations when examining information in quantum systems we will look at some examples. To establish terminology and introduce some basic quantum mechanical concepts, we will first consider a simple one hole experiment where a particle is fired through a hole to a detector screen. The particle, say an electron, is fired from a source S through a hole H in a tungsten screen an observed to strike somewhere on a measurement screen M. Repeated measurements eventually build up a frequency distribution P with respect to the position x that the electron hits the screen.

In quantum mechanics the probability that a particle will leave a source S and strike the screen at some point x is given by the absolute square of a complex number called the *probability amplitude.* Here it is the amplitude that the electron leaving S will strike at x. Using Dirac notation will denote this probability amplitude as:

$$\langle \text{Electron strikes at } x | \text{Electron leaves } S \rangle$$

or more simply:

$$\langle x|S \rangle$$

For particles traveling on segmented routes the probability amplitude of the whole route is given by the product of the amplitudes of each segment. So we have:

$$\langle x|S \rangle = \langle x|H \rangle \langle H|S \rangle$$

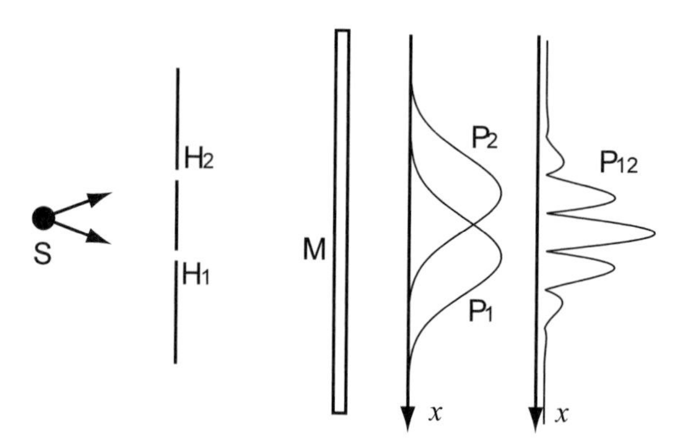

Fig. 4.18. Two Hole Experiment

Now consider a similar experiment using two holes H_1 and H_2 as shown in Fig. 4.18. Here we have two possible paths that could be used to hit the screen at x, through H_1 or H_2. The probability amplitudes of the two routes is given by:

$$\langle x|H_1 \rangle \langle H_1|S \rangle$$

and,

$$\langle x|H_2 \rangle \langle H_2|S \rangle.$$

Note that with two holes the measured frequency is not simply the sum of the two individual distributions P_1 and P_2. It is in fact a more complicated distribution due interference caused by the wave aspect of electron: peaks superposed with peaks reinforce, peaks superposed with troughs cancel. This is roughly shown as P_{12} in Fig. 4.18 and is, as we have just previously defined, $|\langle x|S\rangle|^2$ In determining this probability distribution we must take into account the possibility that the electron may have travelled through either of the two holes. In quantum mechanics, when there is more than one way of an event occurring, then the overall probablity amplitude of the event is the sum of the individual amplitudes of each of the possible ways that the event may have occured. Thus in the two hole experiment the probability amplitude is given by:

$$\langle x|S\rangle = \langle x|H_1\rangle\langle H_1|S\rangle + \langle x|H_2\rangle\langle H_2|S\rangle$$

and thus

$$P_{12} = |\langle x|H_1\rangle\langle H_1|S\rangle + \langle x|H_2\rangle\langle H_2|S\rangle|^2$$

As each probability amplitude is a complex number, distribution we see in P_{12} is generated.

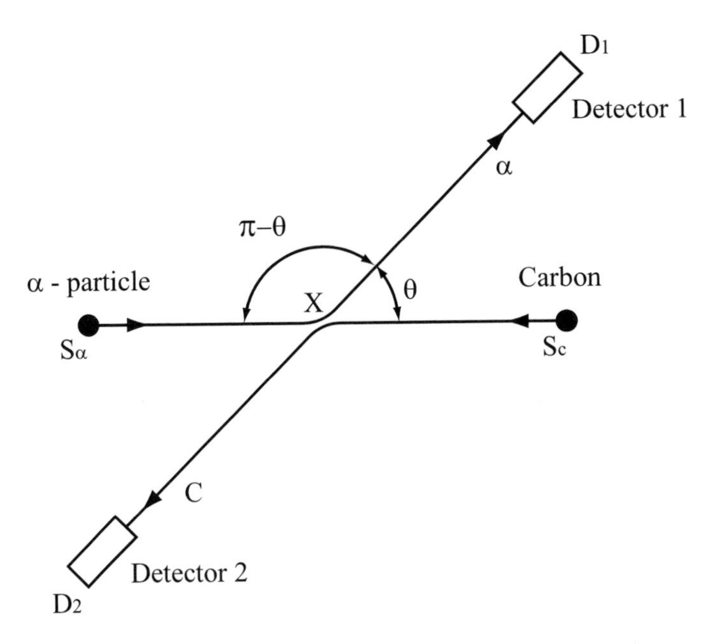

Fig. 4.19. Carbon-Alpha Particle Scattering – θ

Now let us consider a more complicated example. Imagine an experiment in which there is a low-energy collision between alpha particles and carbon atoms causing each to deflect. This scattering experiment is illustrated in Fig. 4.19

α particles are released from a source S_α and carbon atoms are released from source S_C. The α particles and carbon interact at X and scattering occurs in various directions. A detector D_1 is placed to register particles that have been scattered at an angle of θ. Another detector, D_2, is placed directly opposite to D_1 at angle of $(\pi - \theta)$ to the α particle beam. The detectors are capable of detecting both carbon and α particles. Finally, the collisions are assumed to be of low enough energy as to prevent the exchange of nucleons.

The amplitude of a detection of an α particle at D_1 is given by:

$$\langle D_1|X\rangle a\langle x|S_\alpha\rangle$$

where a is the scattering amplitude. There is another possibility of registering a detection at D_1. If the carbon atom is deflected $(\pi - \theta)$ from its course to hit D_1. This possibility is shown in Fig. 4.20

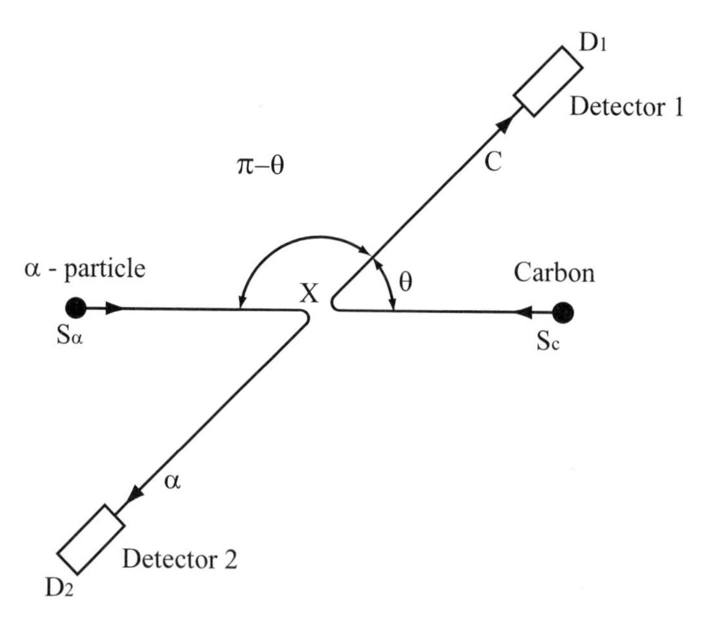

Fig. 4.20. Carbon-Alpha Particle Scattering – $\pi - \theta$

The amplitude of a detection of an α particle at D_1 is given by:

$$\langle D_1|X\rangle b\langle x|S_C\rangle$$

where b is the scattering amplitude.

Now if we conduct the experiment such that a detection event in D_1 is defined as receiving either an α particle *or* a carbon atom without checking which, then we can calculate the probability of detecting some particle in D_1 as:

$$|\langle D_1|X\rangle a\langle x|S_\alpha\rangle|^2 + |\langle D_1|X\rangle b\langle x|S_C\rangle|^2$$

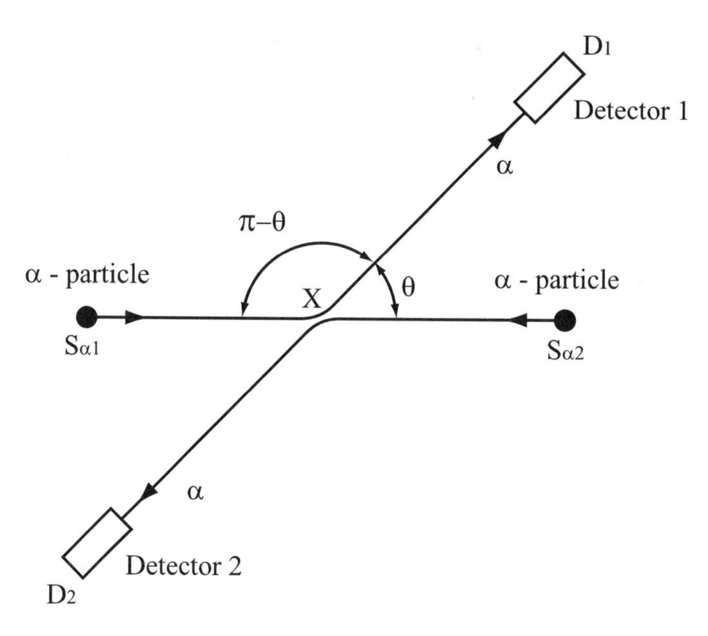

Fig. 4.21. α Particle-αParticle Scattering – θ

This is true whether we check if the α particle or the carbon atom caused the event. This distribution is supported by experimental results. Now, if we conduct the experiment using 2 α particles instead of a carbon atom and an *alpha* particle, then the result we obtain from experimental analysis is quite different. For a θ of 90 degrees, for example, the probability we obtain experimentally is twice that predicted by the above equation. The reason for this is distinguishability. We cannot tell if the α particle we receive at D_1 came from the left source or the right source. There are two possible ways in which the α particle arrived at D_1 scattering θ (Fig. 4.21) or scattering $\pi - \theta$ (Fig. 4.22)

Because the two particles are identical, they are indistinguishable in principle – which was not true of the α particle – carbon atom system.

To calculate the probability of the observation of an α particle at D_1 we must square the sum of the individual possible amplitudes.

Thus the probability of detecting an α particle at D_1 is:

$$|\langle D_1|X\rangle a\langle x|S_{\alpha 1}\rangle + \langle D_1|X\rangle a\langle x|S_{\alpha 2}\rangle|^2$$

It is an important consequence of quantum mechanics that when an event can happen in two indistinguishable ways that there is *always* an interference of probability amplitudes. To further illustrate this notion of quantum distinguishability consider the same experiment using electrons rather than alpha particles. Electrons have spin 1/2 thus they can be in one of two states: spin up or spin down. When two electrons of opposite spin interact there exists the possibility of them both reversing their spins.[25]

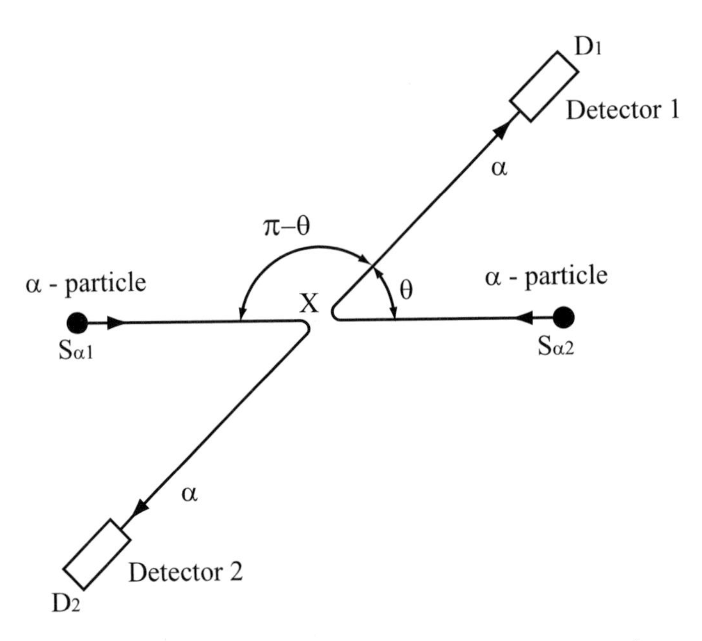

Fig. 4.22. αParticle-αParticle Scattering – $\pi - \theta$

As spin offers a possibility for distinguishing between electrons, the total probablity equation has two components: the indistinguishable amplitudes (where the spins are the same at the sources and also at the detectors), and the distinguishable case where the spins differ. There is a difference however when calculating amplitudes of the indistinguish-

[25] Both must flip spins to conserve angular momentum.

able cases. When protons or electrons interfere the new amplitude interferes with the old with opposite phase, so that the net indistinguishable amplitude is given by

$$|\langle D_1|X\rangle a\langle X|S_{e1}\rangle - \langle D_1|X\rangle a\langle X|S_{e2}\rangle|^2$$

where we designate the two sources to be S_{e1} and S_{e2}. This is true for the scattering of all fermi particles.

For the complete system of all probabilities for our electron scattering experiment, we can generate Table 4.2.[26] The final line in the table gives the total probability of the scattering experiment taking into account interference caused by indistinguishability. As previously

Table 4.2. Probabilities for electron-electron scattering

Fraction of Cases	Spin at S_{e1}	Spin at S_{e2}	Spin at D_1	Spin at D_2	Probability
1/4	up	up	up	up	$\lvert\langle D_1\lvert X\rangle a\langle X\lvert S_{e1}\rangle - \langle D_1\lvert X\rangle a\langle X\lvert S_{e2}\rangle\rvert^2$
1/4	down	down	down	down	$\lvert\langle D_1\lvert X\rangle a\langle X\lvert S_{e1}\rangle - \langle D_1\lvert X\rangle a\langle X\lvert S_{e2}\rangle\rvert^2$
1/8	up	down	up	down	$\lvert\langle D_1\lvert X\rangle a\langle X\lvert S_{e1}\rangle\rvert^2$
1/8	up	down	down	up	$\lvert\langle D_1\lvert X\rangle a\langle X\lvert S_{e2}\rangle\rvert^2$
1/8	down	up	up	down	$\lvert\langle D_1\lvert X\rangle a\langle X\lvert S_{e2}\rangle\rvert^2$
1/8	down	up	down	up	$\lvert\langle D_1\lvert X\rangle a\langle X\lvert S_{e1}\rangle\rvert^2$
Probability					$= \frac{1}{2}\lvert\langle D_1\lvert X\rangle a\langle X\lvert S_{e1}\rangle\rvert - \lvert\langle D_1\lvert X\rangle a\langle X\lvert S_{e2}\rangle\rvert^2 + \frac{1}{2}\lvert\langle D_1\lvert X\rangle a\langle X\lvert S_{e1}\rangle\rvert^2 + \frac{1}{2}\lvert\langle D_1\lvert X\rangle a\langle X\lvert S_{e2}\rangle\rvert^2$

mentioned, the uncertainty principle places a limit on what we can measure in a quantum system and thus what sort of information can be transferred. If, for instance, a photon is prepared in one of two defined nonorthogonal polarized states s_1 or s_2 and then passed through a beam splitter [27] with orientation θ, the probability that a photon in state s_1 will pass through the splitter is $|s_1 \cdot \theta|^2$ and the probability that a photon in state s_2 will pass through the splitter is $|s_2 \cdot \theta|^2$. However since s_1 and s_2 are nonorthogonal there is no orientation of the crystal which will permit the passage of just one and not the other. The

[26] This is based on [34] which offers a more thorough treatment of the matter in this section.

[27] A beam splitter is a crystal with polarized orientation which will allow the passage of a photon or deflect the photon at an angle dependent on the polarization of the photon relative to the orientation of the crystal.

most we could do to extract information regarding the two states is to perform repeated experiments and build up probability distributions. It is these distributions that would have to be used by any IGUS in apprehending information concerning the photon states.

4.8 Symmetries and Algorithmic Information Theory

4.8.1 Symmetry and Kolmogorov Complexity

In its simplest form, the Kolmogorov-Solomonoff-Chaitin theory of information states that the information contained in a numeric string is the length in bits of the smallest program that can be run on a Turing machine to generate that string. Strings, under my group theoretic model, may be shown to contain symmetries. That is, they may contain relationships between their substrings which might be encapsulated in formulae. For example, a string with a repeating substring, such as 12345123451234512345. . . , contains a translation symmetry that may be exploited in a program used to generate the full string. The use of algorithms that are descriptively shorter than their output is an example of the embodiment of symmetries that act on a string.

4.8.2 Memory and Measurement

Brillouin [10] and Szilard [79] demonstrated the importance of measurement and memory in information systems. Measurement is the ability to apprehend distinguishable states of an entity. The interpretation of distinguishability with regards to numeric strings is perhaps not immediately obvious. We have noted that distinguishability is the extrinsic quality of an object which permits one to say that it is one specific entity and not another or that the object is in one particular state and not another. In *Definition 1* we saw that an IGUS can distinguish between entities P and Q just in case P and Q do not possess all properties in common. Distinguishing between two finite strings is a straightforward matter: we simply compare the digits in each corresponding place in the strings to see if they match. It is assumed that this is an error-free process. But how do we compare infinite strings? We can do this computationally by comparing the strings to a finite number of significant figures. Modern computation uses a finite number, 1, 2, 4, 8, 16 . . . , of bytes to store numbers of increasing size or accuracy. This represents string distinguishability *in practice.* All properties of the string are not available to the IGUS. In theoretical work we can compare two infinite strings and determine that they are not the same. This corresponds to being distinguishable *in principle*. Here the two strings possess properties that differ in some respect.

Memory in an IGUS is a capacity necessary for storing state information about an informatic object. Landauer [49] and Bennett [5] have shown that memory (specifically the erasure thereof) has thermodynamic consequences in computational systems. To understand the role memory plays in Algorithmic Information Theory we return to Turing machines. Both Turing and Chaitin considered the use of a scratch-tape. Turing acknowledges the need for "rough notes to 'assist the memory'. It will only be these rough notes which will be liable to erasure" [81]. While Chaitin [21], in defining his Turing machine, proposes the inclusion of a work tape to store intermediate results. Data on Chaitin's work tape can be read, written and erased. Turing and Chaitin both pass over the inclusion of the work tapes as a matter of practical implementation, of no real theoretical import. However we have already seen that the role that memory plays is crucial. In a system where a Turing machine is generating a string, the work tape contains state representations of the asymmetries associated with that string. This will be demonstrated in the next section.

4.8.3 Groups and Algorithmic Information Theory

Kolmogorov's definition of the absolute amount of information in a sequence is the length of the shortest instruction, of all the possible instructions that generates that sequence. The complexity K (information) of a sequence y generated by a program p is given by:

$$K_{\varphi}(y) = \begin{cases} \min_{\varphi(p)=y} l(p) \\ \infty, \text{ If there is no p such that } \varphi(\text{p}) = \text{y} \end{cases}$$

l is a length operator.

Chaitin offers an almost identical definition, where a Turing machine M running a program P generating an output string S

$$L_m(S) = \begin{cases} \min_{M(P)=S}(\text{Length of P}) \\ \infty, \text{ If there is no such P} \end{cases}$$

Both accounts maintain that the information contained in a string is the shortest partially recursive function capable of generating that string or the string itself. If the asymmetry account of information is correct we should be able to show how symmetries are exploited in some partially recursive function to compress information in certain sequences.

In order to see how Group Theory and symmetries fit in this paradigm we shall use a Turing machine M, not to generate a string but to examine actions of functions on a preexisting string. Consider a four tape Turing machine consisting of:

- An input tape, I.
- A program tape, P.
- A work tape, W.
- An output tape, O.

The input tape is read-only and contains a representation of the string under examination. Reading this tape constitutes the measurement step previously mentioned. The nth position of the input tape is designated I_n. The program tape is also read-only and contains the algorithm to be tested. The work tape can be read, written to and erased. It is the memory storage for the system. The nth position of the input tape is designated W_n. The output tape is write-only and contains the results of the algorithm acting on the input string. The nth position of the input tape is designated O_n.

Consider a program tape that contains an appropriate representation [28] of the following set of instructions:

Step 1	Read I_1
Step 2	Write this value to W_1
Step 3	Read I_4
Step 4	Write this value to W_2
Step 5	Read W_1
Step 6	Write this value 3 consecutive times on O starting at the first blank O_n.
Step 7	Read W_2
Step 8	Write this value 3 consecutive times on O starting at the first blank O_n.
Step 9	If halting condition[29] is met, halt.
Step 10	Goto step 5.

[28] By appropriate I mean an encoding of the execution process capable of being executed as discussed in Section 2.2.2.

[29] The halting condition is some predefine criterion for exiting the algorithm. If we are generating a finite length sequence, the condition may be a count of write squares on the input tape. The definition of this step will control whether the machine is a partial recursive function or not. If the criterion is defined in such a way that the machine does not halt for all inputs then it is a partial recursive function.

Lets examine how this system works on an obviously non-random string encoded on the input tape. Lets use the following sequence:

$$777000777000777000\ldots$$

On completion of execution we find that comparing the input and output tape shows that they are indistinguishable. The operation of M is a automorphism acting on an ordered set (a sequence of decimal digits) which left the set invariant. Thus it is a symmetry. The algorithm is obviously contrived to be a symmetry for the above input string, but it also is a symmetry when acting on any of the other 99 possible decimal sequences of the form $xxxyyyxxxyyyxxxyyy\ldots$ If M were to act on sequences of any other form, the input and output tapes would be distinguishable and hence the action would be an asymmetry. Correspondingly if the program, in steps 6 and 8, wrote the digit out 4 times, the action would not be a symmetry for our input string but would be for inputs of the form $xxxxyyyyxxxxyyyyxxxxyyyy\ldots$

We note that in steps 2 and 4, the storing of the unique digits onto the working tape. This is an example of storing asymmetry state details in memory. For another example of a symmetry Turing transformation acting on non-random sequences consider Champernowne's constant in base 10 without the decimal place. The Turing program would look like:

Step 1	Read I_1
Step 2	Write this value to W_1
Step 3	Read W_1
Step 4	Add 1 to value read in step 3 write to next blank O_n
Step 5	Write also to W_1
Step 6	If not halt, goto step 3

Now we consider random strings. If had input tape encoded a purely random sequence and we could not find an algorithm to generate that sequence, then the only Turing machine symmetry that we could generate would be a program as follows:

Step 1	Read I_1
Step 2	Write this value to O_1
Step 3	Read next I_n
Step 4	Write this value to O_n
Step 5	If not the end of input, goto step 3
Step 6	Halt

Here there is no use of memory as no real work is being done. True, the input sequence is indistinguishable from the output sequence but it is just trivial copying. This corresponds to an identity mapping automorphism.

For generating sequences using Turing machines in the Kolmogorov-Chaitin sense one dispenses with the input tape and "hardcodes" the selection procedures into the program.

We are now in a position to demonstrate that the four-tape Turing machines described above can constitute a group under Group Theory.

Lemma 1. *Let G be a set of automorphisms and if each $g_i \in G$ is generated by a Turing machine M_i, consisting of an input tape I, an output tape O, a working tape W and a program tape P, then the set G together with a sequential application multiplier forms a group.*

Proof. G will be shown to be a group if:

1. The multiplication of elements $g_i \in G$ is associative.
2. There exists an identity element, e.
3. Each element in G has an inverse.

The multiplication operator is sequential application of the automorphisms and hence $g_1(g_2 g_3) = (g_1 g_2)g_3$.

There exists an identity element e, namely the program above that directly copies the input tape to the output tape, such that $ge = g = eg$

It has been shown by Bennett[5] that any Turing machine can be logically reversible provide it records all intermediate steps on the working tape. Thus:

$$\forall g \in G \, \exists g^{-1} : gg^{-1} = e.$$

Thus the set G constitutes a group. □

We see that the automorphisms are associative since, as with the application of geometric transforms such as rotations, the multiplication operator is serial application. The group, in most cases, will not be abelian as the order of application will be crucial and thus $AB \neq BA$. The identity transform e is the simple transcription of the input tape to the output tape and in the case of truely random input sequences, G should be a singleton set with e as its only member.

Bennett [5] notes that with the non-erasure of the working tape a Turing machine is logically reversible, but he has shown that by applying 3 stage process, where each stage is reversible, one can generate the output tape, erase the working tape and have a reconstructed copy of the input tape. Kolmogorov himself acknowledges that the theory provides no indication of *how* one is to find the symmetries to arrive at the shortest program. (Just as Jaynes provides guidance other than using "common-sense" on the selection of germane transformations.) It, like the quest for regularities in science as a whole, is a process of inspection, experimentation and discovery. In this manner the Kolmogorov account of information is, in a non-proscriptive way, another manifestation of the action of group transforms on sets to find symmetries (or asymmetries). Here they are represented in an algorithmic fashion. Many trivial functions may be constructed to be symmetries or asymmetries for a given sequence, however functions should only be considered for inclusion if they are relevent to the problem at hand.

4.8.4 Symmetry and Randomness

Algorithmic Information Theory and a group theoretic account of information are linked by notions of redundancy and compression. As Chaitin noted (see Section 2.2.3), random strings are "patternless". There are no redundancies in the sequence that may be exploited in the generation of the string and so the length of a program whichgenerates a finite random string must be approximately the length of the string itself. A string is random just in case it cannot be algorithmically compressed. In informatic terms, this means the total information in the string is the string itself. There are no symmetries that may be exploited to arrive at a reduced information account. A random string is one that is maximally asymmetric. It is another example of Case Maximum Asymmetry as specified in Section 4.2.2.

However there exist strings which appears random yet can be generated from relatively simple algorithms. Consider for example the number π. If it is to be generated by a simple algorithm and thus be highly symmetrical, how is it that the infinite sequence of digits that constitutes the transcendental 'appears' random and has not yet been proved to fail any current tests of randomness? To consider this under our current schema, we must ask where does the symmetry lie? The number π is defined as the ratio of a circle's circumference to its diameter. Let us define C as the set of all circles and $\Re$ the set of real numbers. For any circle $c \in C$ nominate $S_c \in \Re$ to be the circumference of c and $r_c \in \Re$ to be the radius of c. We define a function $f(x, y) = (x, 2\pi x)$. By inspection we can see that $\forall c \in C, f(r_c, S_c) = (r_c, S_c)$. That is, the transform f we have defined is a symmetry for all (r_c, S_c).[30] The function f is a non-trivial function since the substitution of any other real

[30] We could create an infinite number of such symmetries by including trivial functions. However, these should not be included in the information calculation as,

number for π in f would not produce a symmetry. So the symmetric 'work' is being done by the number π; it is in fact a symmetry in this system.

We can manipulate f to return to the original definition of π so that for any S_c and r_c we can determine the numeric representation of π. The notions of circumference and radius are geometric ones so to determine π algebraically, special formulæ are employed. These are iterative methods that converge to π after many iterations, some converging more rapidly than others. These include a special case of the Wallis formula

$$\pi = 2\prod_{n=1}^{\infty}\left[\frac{(2n)^2}{(2n-1)(2n+1)}\right] = 2\frac{2\times 2}{1\times 3}\frac{4\times 4}{3\times 5}\frac{6\times 6}{5\times 7}\cdots$$

and Euler's convergence improvement transformation:

$$\pi = 2\sum_{n=0}^{\infty}\frac{n!}{(2n+1)!!}$$

$$= 2\left(1+\frac{1}{3}+\frac{1\cdot 2}{3\cdot 5}+\frac{1\cdot 2\cdot 3}{3\cdot 5\cdot 7}+\cdots\right).$$

Until recently, the most popular algorithms used to generate π have been variations of Machin's 1709 formula:

$$\frac{\pi}{4} = 4\tan^{-1}\left(\frac{1}{5}\right) - \tan^{-1}\left(\frac{1}{239}\right)$$

However these are slow to converge. For example, Euler's equation only converges at one bit per term. More recent innovations, such the Brent-Salamin Algorithm [69], provide methods to produce π that converge quadratically or faster.

The infinite digital sequence that these algorithms produce has, in decimal notation, the following first 600 digits:

---

from Burnside's Lemma, they are non-germane transformations (like time of day or colour of the circle in Jaynes' straw and circle example in Section 4.4.1).

3.

1415926535897932384626433832795028841971693993751058209749 44

5923078164062862089986280348253421170679821480865132823066 47

0938446095505822317253594081284811174502841027019385211055 59

6446229489549303819644288109756659334461284756482337867831 65

2712019091456485669234603486104543266482133936072602491412 73

7245870066063155881748815209209628292540917153643678925903 60

0113305305488204665213841469519415116094330572703657595919 53

0921861173819326117931051185480744623799627495673518857527 24

8912279381830119491298336733624406566430860213949463952247 37

1907021798609437027705392171762931767523846748184676694051 32

It is widely held that the number π not only has the property of decimal normality, but is absolutely normal, that is, normal in all base systems. However, normality for π has never been proven in *any* base system. Normality, we recall (Section 2.2.3), is a necessary (but not sufficient) test for randomness. Under Löf-Martin's definition of randomness, a string must pass all tests of randomness. Proving the π is normal would be an important step down the path of proving the string is indeed random.

In 1996 a radically different approach for generating π was developed by Bailey, Borwein and Plouffe [2]. The elegant equation is a digit-extraction algorithm[31] which is used to generate identities for numbers such as π, π^2, $\log_e(2)$. A special variant of the algorithm offers π critical identity in the hexadecimal system as follows:

$$\pi = \sum_{i=0}^{\infty} \frac{1}{16^i} \left(\frac{4}{8i+1} - \frac{2}{8i+4} - \frac{1}{8i+5} - \frac{1}{8i+6} \right)$$

This type of formula can also be used to generate π in base 2 but not in decimal.[32] The importance of this approach lies in the fact the one can

[31] A digit-extraction algorithm is one which computes specific digits of a given number without requiring the calculation of previous digits.

[32] Borwein, Borwein & Galway have shown that π has no Machin-type BBP arctangent formula when the base ? 2 and they conjecture that "to the best of our knowledge, when there is no Machin-type BBP formula for a constant then no

compute the d-th digit on the binary π string without calculating any other digits. This strongly implies that the digits of π are independent and this could provide the starting point for proving, first, that π is normal and, second, that it is random. One approach toward achieving these proofs has been to translate the problem of the normality of π into a problem of chaotic dynamics (Bailey and Crandell, 2001). This approach examines sequences generated by the BBP formula against the conjecture that they are uniformly distributed between 0 and 1. If the conjecture is proven, π can be shown to be normal.

The algorithms presented in this section for the generation of the sequence of digits of π characterize different ways of representing the symmetry embodied in π, that is the relationship between the diameter and the circumference of a circle. This symmetry is only relevant to the sets of S and r as previous defined. As previously noted, any real value other than π active in π's stead in the transform f when f acts on (S_c, r_c) does not produce a symmetry. Recalling our discussion of groups in Section 4.2, this is analogous to a Type I rotation[33] of 120 degrees of a tetrahedron producing a symmetry, whereas Type I rotation of 180 degrees of a tetrahedron fails to produce a symmetry. As a corollary, the symmetry is only defined as a symmetry with respect to the set on which it acts. A Type I rotation of 120 degrees of a tetrahedron producing a symmetry, whereas Type I rotation of 120 degrees acting on a cube (see Fig. 4.4) fails to produce a symmetry. To restate: a symmetry is only defined with respect to a system. If π is employed in a function acting on two sets of real numbers other than S and r, there is no guarantee that a symmetry would result.

Thus it is important to realise that one shouldn't confound a symmetry and the set on which it acts. The number π is a symmetry and the set on which it acts is the combined pair of S_c and r_c. As symmetry, π can be represented as an infinite string or it can be represented in a variety of symbolically more compressed ways which describe the relationship between any S_c and r_c (where $c \in C$). If we wish to consider the action π as a symmetry in terms of compression, we must look at the reduction in number of possible translations from the $r \Rightarrow S$ mapping to the specific $r_c \Rightarrow S_c$ mapping. That is which real numbers in S

BBP formula of any form is known for that constant" [9]. You cannot convert a specified binary integer in the binary π string to decimal without knowing all the preceding bits. This rules the method invalid as a digit-extraction algorithm. It doesn't mean, however, that other, non-BBP, digit-extraction algorithms aren't possible in base-10.

[33] A Type I rotation is a rotation around an axis passing through the centroid and a vertex. See Fig. 4.1.

correspond to those in r when S is to be the circumference of the circle whose radius is r.

However it is π represented as a string with automorphisms acting on it that is most intriguing. If the sequence passes all tests of randomness and yet is capable of being generated by relatively simple algorithms then what are we to say about about its informatic content? Is it random or not? Work is still being conducted on the sequence and others like e and perhaps a rethink of the notions of compressibility and randomness will be warranted if these sequences are found to be random. However it is still clear that in terms of the transfer of information these sequences are capable of being substantially compressed implying a great deal of redundancy.

In this section we have seen that where strings are compressible, where there are redundancies in strings, symmetries exist that may be exploited by algorithms to produce more condensed informatic descriptors. Conversely we have seen that a string is random just in case it cannot be algorithmically compressed; there are no symmetries that may be used to arrived at a reduced information account. A random string is one that is maximally asymmetric. It is the concepts of compressibility, randomness and symmetry that wed the Asymmetry Principle of Information to Algorithmic Information Theory.

4.8.5 A Final Signpost

In this chapter we have examined the relationship between symmetry and information and determined it to be such that asymmetry underlies the concept of information. That is, information is a way of abstractly expressing asymmetries. We have seen that information can be quantified using the algebra of symmetry – Group Theory – and that this quantitative method has direct application to real-world problems. Jaynes' Principle of Maximum Entropy has been demonstrated to be a technique of incorporating transformations that account for asymmetries in a model or theory to arrive at a truly assumption-free prior and, though grounded in a subjectivist tradition, still has great relevance as a corollary to the Asymmetry Principle of Information.

The concept of distinguishability, and by extension, asymmetry, has been shown to lie at the core of theories of entropy. By taking symmetries into account in the examination of monatomic, idealized gases, for example, observed entropies can be reconciled with Boltzmann-Maxwell theoretic entropies, leading to the Sackur-Tetrode equation. The relationship between entropy and information, as documented in Section 2.2.1, completes the triangle. Analysis of Maxwellian Demons and of the Gibbs' Paradox has revealed the mechanisms by which informatic processes such as sorting have the capacity to perform thermodynamic work. We have also seen that by considering physical entropy examples we find material manifestation of the inverse relationship between symmetry and entropy.

Finally, we have just seen in Algorithmic Information Theory the notions of incompressibility, redundancy and randomness tie asymmetry to algorithmic information.

5

Conclusion

The three-word postulate of this book is: "Information is Asymmetry". It is under this banner that I have attempted to unify the three approaches: the Thermodynamic/Statistical Mechanics Account, Communication Theory and Algorithmic Information Theory. This assertion is, of course, an over-simplification and there are, as always, caveats. I do not maintain that 'asymmetry' and 'information' have an identity relationship. The relationship that they do have needs to be framed in the context of the IGUS-Information Object Schema and must be filtered by the concept of indistinguishability. Nonetheless it is the broad concepts of symmetry and asymmetry that underlie the current theories of information.

There are several aspects of my method that are original and others that extend previous work. The development of distinguishability leading to my production of the IGUS/Information Object model illustration is novel. The formulation of information in terms of group orbits using Burnside's Lemma:

$$\mathrm{I} = \log(\sum_{g \in G} |S^g|) - \log(|G|)$$

is, to the best of my knowledge, a new account. I believe it to be a powerful tool with potential for application in a vast number of fields.

Some of the work presented here is an extension of E.T. Jaynes' work on prior probabilities [38], [39]) and his Maximum Entropy Principle [40],[37]). I have avoided the epistemological position that Jaynes took as I believe that information is an objective quantity capable of relational representation unde the IGUS-object formulation. As an objective quantity, Information is capable of producing work as shown by our consideration of Maxwell's Demon and Gibbs' paradox. Distinction can be made (as in sorting) to create differentials that may be used to generate work. Measurement and memory (at least one binary register) are necessary to produce work and that in systems that are employed cyclically, the memory must be reset after each cycle. It is the resetting of the memory that prevents the violation of the Second Law.

The Information/Asymmetry principle, when applied to the field of Algorithmic Information Theory, has been shown to account for the notions of compressibility, randomness and complexity with respect to algorithms and strings. This is fundamentally due to symmetries creating redundancies in these objects.

The potential for symmetry considerations in Information Theory is vast and this work has only just scratched the surface. The need for further work is great and opportunities plentiful. The approach adopted here is an instance of Representation Theory: a study of the ways that a given group may act on vector spaces. Group Theory and Representation Theory are rich and powerful branches of mathematics. Algebraic structures abound and many problems related to information theory may be posed: Are all group representations of informatic objects necessarily topological groups? What special role do Lie algebras play with relation to informatic objects? Is there a special application for fields? We have concentrated predominately on finite groups because of tractability and Burnside's lemma is applicable *only* to finite groups. Can we handle infinite groups under the current formulation, and if so how?

Beyond mathematics there are applications in metaphysics, particularly causation. It has been proposed that a causal process involves the transfer of a non-zero value conserved quantity [31] and that a token of a particular quantity of information fills this role [25]. I will not debate the hypothesis that "causation is the transfer of information" here but will briefly examine, without bringing in too much additional research at this late juncture, what a group-theoretic approach to information could bring to the table if the hypothesis is assumed to be true.

Stewart and Golubitsky [78] consider causal processes in which symmetry is broken. For example, consider a perfectly spherical droplet of milk falling into a still bowl of the same liquid. Initially the droplet has an $O(3)$ symmetry group. On impact, three dimensional symmetry is lost and a thin-walled ring rises. At this point we have a subgroup, $O(2)$. of symmetries. As the ring of milk rises it curves outward until the continuous circle is broken and regular spikes are thrown up giving the appearance of a crown. The $O(2)$ symmetry is broken in that rotational symmetry of all angles are no longer possible, just a subset of those defined by the angular distance between the spikes in the crown. The group has gone from O(2) to a dihedral subgroup D_{24}. Steward and Golubitsky maintain that the physical outcome of the causal process is just one of many symmetrically related possible outcomes. The milk crown may not have appeared at exactly the same rotation as it did. If the experiment were repeated we would get a D_{24} crown but it might be rotated somewhat. All the possible manifestations of the D_{24} crown are related to each other in a way defined by the O(2) group. In this way symmetry breaking in the causal process is a form of resolution of the possible into the actual.

Another example is the deformation of a perfect sphere under an axial load (think of a ping-pong ball buckling under uniform radial force). A circular dent develops around the force axis breaking the O(3) symmetry but retaining the circular O(2) symmetry.

To capture this, Stewart and Golubitsky extend Pierre Curie's Symmetry Principle[1] to give the following definition:

> "Extended Curie Principle: physically realizable states of a symmetric system come in bunches, related to each other by symmetry. To put it another way, a symmetric cause produces one from a symmetrically related set of effects. The Extended Curie Principle isn't quite as simple or elegant as the original version, but it has the advantage of being correct" [78].

Joe Rosen offers an ostensibly diametrically opposed position. In the formulation of his Symmetry Principle he states, "The symmetry group of the cause is a subgroup of the symmetry group of the effect. Or less precisely: The effect is at least as symmetric as the cause" [68].

The justification for this is that there are many states that could effectively constitute a cause in terms of a law-like account of a causal system. These are known as *equivalent causes.* Similarly there are many states that constitute an effect: *equivalent effects.* The causation statement is then 'Equivalent causes lead to equivalent effects'.

> "Since cause-equivalence implies effect-equivalence, it follows that every element in the symmetry group of cause must necessarily also be an element of the symmetry group of the effect. There might, of course, be symmetry transformations of the effect that are not also symmetry transformations of the cause" [68].

[1] Curie's principle states that if certain causes produce certain effects, then the symmetries of the causes reappear in the effects produced [28].

I will not examine the veracity of Rosen's claim but more work needs to be done to see a) if these views need to be reconciled, and if so, b) how they can be reconciled. A causal process which is symmetry breaking is one that creates more distinguishing attributes, one that becomes more complex, one that is more informed. A process that creates symmetry in the world – and these exist, at least locally, for someone is making ping-pong balls – reduces the information carrying capacity of the physical set on which it operates.

My feeling is that there are both types of local causal process: information generating and information destroying. But, in accordance with the decree of the Second Law of Thermodynamics, the universe on the whole is becoming more informed. From the initial "perfect symmetry" at the origin of the universe to the formation of matter to complex biological structures, symmetry breaking has proceeded. Potentiality is resolved into actuality and with Group Theory we have a means to account for it. If causation is the transfer of information or, more accurately, the transfer, augmentation and attenuation of symmetry groups, then the benefits could be great indeed; for what we stand to gain is an algebra of causation.

A

Burnside's Lemma

Burnside's Lemma (also known as the Cauchy-Frobenius Lemma[1]) states:

Lemma 2. *Whenever G is a finite group acting on a finite set S, the number of orbits, i.e. distinct configurations of S relative to G, is*

$$O = \frac{1}{|G|} \sum_{g \in G} |S^g|$$

where $|G|$ is the order of the group G and $|S^g|$ is the order of the subset of points $s \in g$ fixed by g, that is $g(s) = s$.

Proof. The elements of S are partitioned into their orbits under G. We desire the total number of orbits. If a finite set S is partitioned into orbits, then

[1] The Lemma was indeed first stated, in an initial form, by Cauchy [17] and later by Frobenius [35]. It was rediscovered by Burnside in 1900 [13]. It was eventually extended by Pólya [64] for application in combinatorial counting problems.

$$\begin{aligned}
\text{Number of Orbits} &= \sum_{s\in S} \frac{1}{(\text{Orbit size of } s)} \\
\sum_{g\in G} |S^g| &= \sum_{s\in S} (\text{Number of } g \text{ which fix } s) \\
&= \sum_{s\in S} \frac{|G|}{(\text{Orbit size of } s)} \\
&= |G| \sum_{s\in S} \frac{1}{(\text{Orbit size of s})} \\
\frac{1}{|G|} \sum_{g\in G} |S^g| &= \sum_{s\in S} \frac{1}{(\text{Orbit size of } s)} \\
&= \text{Number of Orbits } = O \quad \square
\end{aligned}$$

B

Worked Examples

B.1 Clocks

B.1.1 Case 1

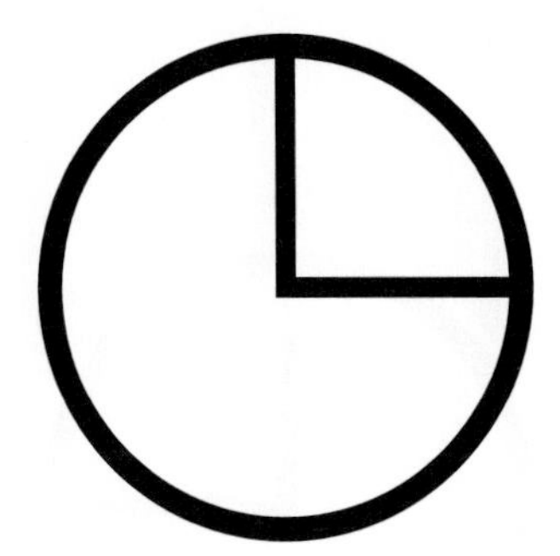

Fig. B.1. High Symmetry Clock

Assumptions:

1. The clock has only one face (that is it is not reversible);
   a) The minute hand is always pointing at a minute divisor (angular separation of 6 degrees);
   b) The minute hand moves between the minute divisors infinitely quickly. Thus the minute hand can be in one of 60 possible states;
   c) The hour hand is always pointing at an hour divisor (angular separation of 30 degrees);
   d) The hour hand moves between the divisors infinitely quickly on the change of the hour. The hour hand can be in one of 12 possible states.

e) Both hands are identical

In total there are 720 formal states, $s \in S$. We will use the following notation hh:mm (e.g. 5:14) in the following notes to refer to the position of the hour hand (hh) and the minute hand(mm).

Apart from the identity transform there are two classes of symmetry: reflection (due to the hands being identical) and rotation (due to lack of reference on the face). Let φ_x denote reflection about an axis rotated x degrees around the centre. Let r_y denote rotation of y degrees around centre.

We have then,

$$\sum_{g \in G} |S^g| = 2448$$

Using Burnside's lemma, the number of orbits is:

$$\frac{1}{|G|} \sum_{g \in G} |S^g| = \frac{2448}{24} = 102$$

B.1.2 Case 2

Fig. B.2. Medium Symmetry Clock

Assumptions:

1. The clock has only one face (that is it is not reversible);
2. The minute hand is always pointing at a minute divisor (angular separation of 6 degrees);
3. The minute hand moves between the minute divisors infinitely quickly. Thus the minute hand can be in one of 60 possible states;
4. The hour hand is always pointing at an hour divisor (angular separation of 30 degrees);

Possible transforms $g \in G$	Number of elements of $s \in S$ fixed by g	Notes
e	720	All possible s are fixed by the identity transform
φ_0	12	Such as 0:0, 1:55, 2:50, 3:45, 4:40, 5:35, 6:30, 7:25, 8:20, 9:15, 10:10, 11:05
φ_{15}	12	Such as 0:05, 11:10, 10: 15, 9:20, 8:25, 7:30, 6:35, 5:40, 4:45, 3:50, 2:55, 1:00
φ_{30}	12	Similar to φ_0
φ_{45}	12	Similar to φ_{15}
φ_{60}	12	Similar to φ_0
φ_{75}	12	Similar to φ_{15}
φ_{90}	12	Similar to φ_0
φ_{105}	12	Similar to φ_{15}
φ_{120}	12	Similar to φ_0
φ_{135}	12	Similar to φ_{15}
φ_{150}	12	Similar to φ_0
φ_{165}	12	Similar to φ_{15}
r_{30}	144	Can't distinguish between 1:10 and 2:15 for example or 4:50 and 5:55
r_{60}	144	Similar to r_{30}
r_{90}	144	Similar to r_{30}
r_{120}	144	Similar to r_{30}
r_{150}	144	Similar to r_{30}
r_{180}	144	Similar to r_{30}
r_{210}	144	Similar to r_{30}
r_{240}	144	Similar to r_{30}
r_{270}	144	Similar to r_{30}
r_{300}	144	Similar to r_{30}
r_{330}	144	Similar to r_{30}
Total	2448	

5. The hour hand moves between the divisors infinitely quickly on the change of the hour. The hour hand can be in one of 12 possible states.
6. The hour hand is shorter than the minute hand.

In total there are 720 formal states, $s \in S$. Again we will use hh:mm notation.

With the removal of the identical hands constraint, the reflective symmetry has been broken. Thus beyond the identity transform there is just the rotational symmetry. Let r_y denote rotation of y degrees around the centre.

Possible transforms $g \in G$	Number of elements of $s \in S$ fixed by g	Notes
e	720	All possible s are fixed by the identity transform
r_{30}	144	Can't distinguish between 1:10 and 2:15 for example or 4:50 and 5:55
r_{60}	144	Similar to r_{30}
r_{90}	144	Similar to r_{30}
r_{120}	144	Similar to r_{30}
r_{150}	144	Similar to r_{30}
r_{180}	144	Similar to r_{30}
r_{210}	144	Similar to r_{30}
r_{240}	144	Similar to r_{30}
r_{270}	144	Similar to r_{30}
r_{300}	144	Similar to r_{30}
r_{330}	144	Similar to r_{30}
Total	2304	

We have then,

$$\sum_{g \in G} |S^g| = 2304$$

Using Burnside's lemma, the number of orbits is:

$$\frac{1}{|G|} \sum_{g \in G} |S^g| = \frac{2304}{12} = 192$$

B.1.3 Case 3

Assumptions:

1. The clock has only one face (that is it is not reversible);
2. The minute hand is always pointing at a minute divisor (angular separation of 6 degrees);
3. The minute hand moves between the minute divisors infinitely quickly. Thus the minute hand can be in one of 60 possible states;
4. The hour hand is always pointing at an hour divisor (angular separation of 30 degrees);

Fig. B.3. Low Symmetry Clock

5. The hour hand moves between the divisors infinitely quickly on the change of the hour. The hour hand can be in one of 12 possible states.
6. The hour hand is shorter than the minute hand.
7. The clock is marked with a symbol (e.g. with a '12' in the above illustration) which must always be at the top.

In total there are 720 formal states. Again we will use hh:mm notation.

The inclusion of the '12' prevents the clock being rotated so now 1:10 and 2:15, for example, can be distinguished. Thus the only remaining transform is the identity transform that fixes all 720 states. So trivially,

$$\sum_{g \in G} |S^g| = 720,$$

and, the number of orbits is:

$$\frac{1}{|G|} \sum_{g \in G} |S^g| = \frac{720}{1} = 720.$$

B.2 Binary String

Consider a 10 place binary string, eg: 0110100111. A ten-place string can be used to represent integers 0 to 1023. Now consider positional translation transforms modulo 10. It may useful to think of as rotational cycling through the 10 bits or as an infinite repetition of the string, for example:

0100111011010011101101001110110100111011010011101101001110 1....

Let $G = \{e, r_1, r_2, r_3, r_4, r_5, r_6, r_7, r_8, r_9\}$ where e is the identity transform, r_x is translation x places modulo 10. Let the group have a sequential "multiplier", M, and let S be the set of all possible 10 bit binary strings.

Transform $g \in G$	Number of elements of $s \in S$ fixed by g	Notes
e	1024	All possible s are fixed by the identity transform
r_1	2	Just 0000000000 and 1111111111 fixed
r_2	4	Just 0000000000, 1111111111, 0101010101 and 1010101010 are fixed
r_3	2	As with r_1
r_4	4	As with r_2
r_5	32	2^5 strings are fixed
r_6	4	As with r_2
r_7	2	As with r_1
r_8	4	As with r_2
r_9	2	As with r_1
Total	1080	

We have then,

$$\sum_{g \in G} |S^g| = 1080$$

Using Burnside's lemma, the number of orbits is:

$$\frac{1}{|G|} \sum_{g \in G} |S^g| = \frac{1080}{10} = 108$$

References

1. F.W. Aston,. *Mass Spectra and Isotopes*, 2nd edn, (Edward Arnold, London, 1942).
2. D. Bailey,P. Borwein, & S. Plouffe, 'On The Rapid Computation Of Various Polylogarithmic Constants', *Mathematics of Computation*, vol. 66, no. 218, pp. 903–913, (1997).
3. D.H. Bailey & R.E.Crandall, R.E., 'On the random character of fundamental constant expansions', *Experimental Mathematics*, vol. 10, June, pp.175–190, (2001).
4. T. Bayes, 'An Essay towards solving a Problem in the Doctrine of Chances', *Phil. Trans. Roy. Soc.*, pp. 370-418, (1763).
5. C.H. Bennett, 'Logical Reversibility of Computation'. In *Maxwell's Demon: Entropy, Information and Computing*, H.S. Leff, & A.F. Rex, (eds) (Adam Hilger, 1990), Bristol, pp. 197–204.
6. C.H. Bennett, 'The Thermodynamics of Computation – a Review'. In *Maxwell's Demon: Entropy, Information and Computing*, H.S. Leff, & A.F. Rex, (eds) (Adam Hilger, 1990), Bristol, pp. 213–248.
7. C.H. Bennett, 'Logical Depth and Physical Complexity'. In *The Universal Turing Machine. A Half Century Survey*, R. Herken. (ed.), 2nd edn., (Springer Verlag, New York, 1995) pp. 207–236.
8. J. Bertrand, *Calcul des probabilités*, (Gauthier-Villars, Paris, 1889), pp. 4–5. Cited in [40], p.477.
9. D. Borwein, J.M. Borwein, & W.F. Galway, 'Finding and Excluding b-ary Machin-Type BBP Formulae', *Canadian J. Math*, January [CECM Preprint 2003] p.195, (2003).
10. L. Brillouin, 'Maxwell's Demon Cannot Operate: Information and Entropy'. In *Maxwell's Demon: Entropy, Information and Computing*, H.S. Leff, & A.F. Rex, (eds) (Adam Hilger, Bristol, 1990) pp. 134–137.
11. L. Brillouin, *Science and Information Theory*, (Academic Press, New York, 1962).

12. E. Brugnoli, & G.D. Farqhar, 2000, 'Photosynthetic Fractionation of Carbon Isotopes'. In *Photosynthesis: Physiology and Metabolism*, R.C. Leegood, T. D. Sharkey, and S. von Caemmerer (eds), (Kluwer Academic Publishers, 2000), pp.399–434.
13. W. Burnside, 'On some properties of groups of odd order', *Proc. Lond. Math. Soc.*, vol. 1, ser. 33, pp. 162–185, (1901).
14. C. Calude, *Information and Randomness: An Algorithmic Perspective*, (Springer-Verlag, Berlin, 1995).
15. R, Carnap, 'Testability and Meaning', *Philosophy of Science*, vol. 3, pp. 419–471, (1936).
16. R. Carnap, *Logical Foundations of Probability*, (University of Chicago Press 1950).
17. A. Cauchy, 'Mémoire sur diverses propriétés remarquables des substitutions régulières ou irrégulières, et des systémes de substitutiones conjugées', *C. R. Acad. Sci. Paris*, vol. 21, 835, Reprinted in *Euvres Complètes d'Augustin Cauchy, Tome IX.* 1896, (Gauthier-Villars, Paris, 1845), pp. 342–360.
18. G.J Chaitin, 1966, 'On the length of programs for computing finite binary sequences', *Journal of the ACM*, 13, pp. 547–569, reprinted in *Information, Randomness and Incompleteness – Papers on Algorithmic Information Theory*, G.J. Chaitin, (World Scientific, 1987), pp. 213–238.
19. G.J Chaitin, 1975, 'Randomness and Mathematical Proof', *Scientific American*, vol. 232 May, pp. 47–52 reprinted in *Information, Randomness and Incompleteness – Papers on Algorithmic Information Theory*, G.J Chaitin,(World Scientific, 1987), pp. 3 – 13.
20. G.J Chaitin, 1969, 'On the Length of Programs for Computing finite binary sequences: Statistical Considerations', *Journal of the ACM*, vol. 16, pp. 145–159 reprinted in *Information, Randomness and Incompleteness – Papers on Algorithmic Information Theory*, G. J. Chaitin,(World Scientific, 1987) pp. 239 –255.
21. G.J Chaitin, 1977, 'Algorithmic Information Theory', *IBM Journal of Research and Development,* vol. 21 pp. 350–359, reprinted in *Information, Randomness and Incompleteness – Papers on Algorithmic Information Theory*, G. J. Chaitin,(World Scientific, 1987), pp. 38–52.
22. R. Clausius, 'Uber verschiedene für die Anwendung bequeme Formen der Hauptgleichungen der mechanischen Wärmetheorie', *Annalen der Physik und Chemie*, vol. 125, pp.353–363, (1865).
23. J. Collier, 'Two faces of Maxwell's Demon Reveal the Nature of Irreversibility', *Studies in the History and Philosophy of Science*, vol. 21, pp. 257–268, (1990).
24. J. Collier, 'Information Originates in Symmetry Breaking', *Symmetry: Science and Culture*, vol. 7, pp.247–256, (1996).
25. J. Collier, 'Causation is the Transfer of Information'. In *Causation, Natural Laws and Explanation*, H. Sankey, (ed.) (Kluwer, 1999) pp. 279–331.

26. A.H. Copeland "Independent Event Histories", *Am. J. of Math*, 51, pp. 612–618, (1929).
27. T.M. Cover, & J.A. Thomas, *Elements of Information Theory.* (John Wiley and Sons, New York. 1991).
28. P. Curie, 'Sur la symerie dans les phenomenes physiques, symerie d'un champ electrique et d'un champ magnetique', *J. de Phys.*, vol. 3, (1894), pp.393–415, (1894).
29. D. Dennett, *Content and Consciousness*, (Routledge & Kegan Paul, London, 1969).
30. M. Dixon & E. C. Webb, *Enzymes*, 3rd edn, (Longman Group, London, 1979).
31. Dowe, P. 'Wesley Salmon's Process Theory of Causality and the Conserved Quantity Theory', *Philosophy of Science*, vol. 59, pp. 195–216, (1992).
32. P. Ehrenfest & T. Ehrenfest *The Conceptual Foundations of the Statistical Approach in Mechanics*, trans. Moravesik, M., (Dover, New York 1959).
33. R.P. Feynman *The Feynman Lectures on Physics*, vol. 1, (Addison-Wesley, 1963) pp.44–46.
34. R.P. Feynman *The Feynman Lectures on Physics*, vol. 3, (Addison-Wesley, 1963) pp. 3–12.
35. F.G. Frobenius, 'Über die Congruenz nach einem aus zwei endlichen Gruppen gebildeten Doppelmodul', *J. reine angew. Math.*, vol. 101, 273–299, (1887). In *Ferdinand Georg Frobenius Gesammelte Abhandlungen, Band II.*, (Springer-Verlag, Berlin, 1968) pp. 304–330.
36. J. W. Gibbs,*Elementary Principles in Statistical Mechanics*, Reprint (Ox Bow Press, Woodbridge, 1981).
37. E.T. Jaynes, 'Information Theory and Statistical Mechanics'. In *Statistical Physics,* Ford, K. (ed.), (Benjamin, New York, 1963) p. 181.
38. E.T. Jaynes,*Prior Probabilities and Transformation Groups*, (1965) unpublished manuscript available from http://bayes.wustl.edu/etj/articles/groups.pdf.
39. E.T. Jaynes, 'Prior Probabilities', *IEEE Transactions On System Science and Cybernetics*, vol. 4, no. 3, pp. 227–241, (1968).
40. E.T. Jaynes, 'The Well Posed Problem', *Foundations of Physics*, vol. 3, pp. 477–493, (1973).
41. E.T. Jaynes, 'Where do we stand on Maximum Entropy?'. In *The Maximum Entropy Formalism*, Levine, R.D. & Tribus, M. (eds), (MIT Press, 1978) pp.15–118.
42. E.T. Jaynes, 'The Gibbs Paradox'. In *Maximum Entropy and Bayesian Methods*, Smith, C.R., Erikson, G.J and Neudorfer, P.O. (eds), (Kluwer Academic Publishers, Dortrecht, Holland, 1992), pp. 1–22.
43. E.T. Jaynes,*Probability Theory: The Logic of Science*, (1994) unpublished manuscript available from http://bayes.wustl.edu/etj/prob/book.pdf.

44. H. Jeffreys,*Theory of Probability,* (1939) reprinted (Oxford Classic Texts, Oxford Press, Oxford, 1998).
45. G. Jumarie,*Maximum Entropy, Information Without Probability and Complex Fractals: Classical and Quantum Approach,* (Kluwer Academic Publishers, 2000).
46. A.N.Kolmogorov,*Grundbegriffe der Wahrscheinlichkeitsrechnung.* (Springer-Verlag 1933), trans. Morrison, N.: *Foundations of the Theory of Probability,* (AMS Chelsea Publishing 1956).
47. A.N.Kolmogorov, 'On Tables of Random Numbers', *Sankhya: The Indian Journal of Statistics,* Series A, Vol. 25 Part 4, (1963) reprinted in *Theoretical Computer Science* 207, pp. 387–395, (1998).
48. A.N.Kolmogorov, 'Three Approaches to the Quantitative Definition of Information' *Problemy Peredachi Informatsii,* Vol. 1, No 1, pp.3–11, (1965).
49. R. Landauer, 'Irreversibility and Heat Generation in the Computing Process', In *Maxwell's Demon: Entropy, Information and Computing,* H.S. Leff, & A.F. Rex, (eds) (Adam Hilger, 1990), Bristol, pp 188–196.
50. P.S. Marquis de Laplace,, *A Philosophical Essay on Probabilities,* trans. Truscott, F.W. & Emory, F.L., (Dover Publications, New York 1951).
51. H.S. Leff, & A.F. Rex, (eds), *Maxwell's Demon: Entropy, Information and Computing,* (Adam Hilger, Bristol 1990).
52. G.W. Leibniz, *The Monadology and Other Philosophical Writing,* trans R. Latta, R., (Garland, New York 1985).
53. G.W. Leibniz, *The Leibniz-Clarke Correspondence,* Alexander, H.G. (ed.), (Manchester University Press, 1956).
54. M. Li, & P. Vitányi, 'Kolmogorov Complexity and its Applications', *Handbook of Theoretical Computer Science,* van Leeuwen, J. (ed), (Elsevier Science Publishers, 1990) pp.187–254.
55. M. Li, & P. Vitányi, *An Introduction to Kolmogorov Complexity and Its Applications,* (Springer, 1997).
56. E. Mach, *Science of Mechanics,* (1902) p. 395 cited in [80], p. 357.
57. P. Martin-Löf, 'The Definition of Random Sequences', *Information and Control,* vol. 9, pp. 602–619, (1966).
58. R. von Mises, *Probability, Statistic and Truth,* (Dover Publications, New York, 1957).
59. R. von Mises in Marbe's 'Gleichförmigkeit in der Welt und die Wahrscheinlichkeitsrechung', *Die Naturwissenschaften,* Vol. 7, No. 11, pp168–175; No.12, pp.186–192; No. 13, pp. 205–209, (1919).
60. R.K. Murray, D.K. Granner, P.A.. Mayes & V.W. Rodwell, *Harper's Biochemistry,* 24th edn., (Appleton and Lange, 1996).
61. Neumann, P.M., Stoy, G.A. and Thompson, E.C. *Groups and Geometry,* (Oxford University Press, Oxford, 1994).
62. J. Neyman, 'Outline of a Theory of Statistical Estimation based on the Classical Theory of Probability', *Phil. Trans. A.,* vol. 236, pp. 333–380, (1937).

63. M. Planck, *Treatise on Thermodynamics*, 3rd ed. Trans. A. Ogg, (Constable, London, 1926).
64. G. Pólya, 'Kombinatorische Anzahlbestimmungen für Gruppen, Graphen, und chemische Verbindungen', *Acta Math.*, vol. 68, pp. 145–254, (1937).
65. L. S. Pontryagin, *Topological Groups*, (Gordon and Breach, New York, 1966).
66. K.R. Popper, *The Logic of Scientific Discovery*, Routledge Classics (2002), New York.
67. K.R. Popper, *Conjectures and Refutations*, 2nd ed., (Harper Torchbooks, New York 1965).
68. J. Rosen, *Symmetry in Science An Introduction to the General Theory*, (Springer, 1995).
69. E. Salamin, 'Computation of π using Arithmetic-Geometric Mean', *Math. Comput.*, vol. 30, pp. 565–570, (1976).
70. E. Schrödinger, 'What is Life?' in *What is Life? with Mind and Matter and Autobiographical Sketches*, (Cambridge University Press, Canto Edition, Cambridge. 2000).
71. E. Schrödinger, *Statistical Thermodynamics* (Cambridge University Press, 1952).
72. C.E. Shannon, 'A Mathematical Theory of Communication', *The Bell System Technical Journal*, vol. 27, no. 3. (1948)
73. Sklar, L. *Physics and Chance*, (Cambridge University Press, Cambridge, 1993).
74. M. v. Smoluchowski, 'Experimentell nachweisbare der üblichen Thermodynamik widersprechende Molekularphänomene,' *Physik Z.* 13, pp. 1069–1080, (1912).
75. M. v. Smoluchowski, 'Gültigkeitsgrenzen des zweiten Hauptsatzes der Wärmtheorie.' *Vorträge über die Kinetische Theorie der Materie und der Elektrizität* (Teubner, Leipzig, 1914), pp. 89–121.
76. R. Solomonoff, *A preliminary report on a general theory of inductive inference.* Technical Report ZTB-138, (Zator Company, Cambridge, Mass. 1960).
77. R.M. Stephenson, & S. Malanowski, *Handbook of the Thermodynamics of Organic Compounds*, (Elsevier, New York, 1987).
78. I. Stewart, & M. Golubitsky, *Fearful Symmetry? Is God a Geometer?*, (Blackwell, Cambridge, Mass. 1992).
79. L. Szilard, 'On the Decrease of Entropy in a Thermodynamic System by the Intervention of Intelligent Beings', (1923) trans. A. Rapaport, & M. Knoller. In *Maxwell's Demon: Entropy, Information and Computing*, H.S. Leff, & A.F. Rex, (eds) (Adam Hilger, Bristol 1990), pp 124–133.
80. D.W. Thompson, *On Growth and Form*, (Dover Publications New York, 1992).

81. A. Turing, 'On Computable Numbers, with an application to the Entscheidungsproblem', *P. Lond. Math. Soc.*, vol. 2, no. 42, pp. 230–265, (1937).
82. J. Ville *Etude Critique de la Notion de Collectif*, (Gauthier-Villars, 1939).
83. A. Wald *Ergebnisse eines Mathematischen Kolloquiums*, vol.8, pp. 38–72, (1937).
84. F.T. Wall, *Chemical Thermodynamics*, (W.H. Freeman and Company, San Francisco, 1958).
85. J. Wallis, *Arithmetica Infinitorum.* (Oxford, 1656).
86. E.W. Weisstein, textitCRC Concise Encyclopedia of Mathematics, 2nd edn, (Chapman and Hall/CRC, 2002).
87. H. Weyl, *Symmetry*, (Princeton University Press. Princeton, 1952).
88. M.W. Zemansky & R.H. Dittman, 1997, *Heat and Thermodynamics*, 7th edn., McGraw-Hill.
89. W.H. Zurek, 1990, 'Algorithmic Information Content, Church Turing Thesis, Physical Entropy and Maxwell's Demon'. In *Complexity, Entropy and the Physics of Information*, W.H. Zurek (ed.), Addison-Wesley, pp.73–89.

Index

Printing: Krips bv, Meppel
Binding: Stürtz, Würzburg

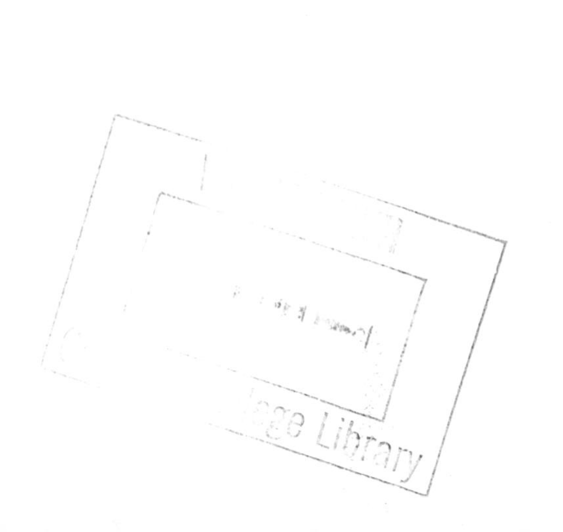